Hydrogen Energetics

Hydrogen Energetics

A Future Clean Energy Carrier

Roman J. Press
NY, USA

WILEY

This edition first published 2025

Registered Offices
John Wiley & Sons, Inc., 111 River Street, Hoboken, NJ 07030, USA
John Wiley & Sons Ltd, The Atrium, Southern Gate, Chichester, West Sussex, PO19 8SQ, UK

For details of our global editorial offices, customer services, and more information about Wiley products visit us at www.wiley.com.

Wiley also publishes its books in a variety of electronic formats and by print-on-demand. Some content that appears in standard print versions of this book may not be available in other formats.

Library of Congress Cataloging-in-Publication Data Applied for:

Hardback ISBN: 9781394173280

Cover Design: Wiley
Cover Image: © scharfsinn86/Adobe Stock Photos

Set in 9.5/12.5pt STIXTwoText by Straive, Pondicherry, India
Printed and bound by CPI Group (UK) Ltd, Croydon, CR0 4YY

C9781394173280_031024

Contents

Preface

The relentless march of civilization has historically been fueled by the consumption of hydrocarbon resources, a practice increasingly linked to climate change and health issues. The announcement of a breakthrough in nuclear fusion energy on 12.13.22 marked a pivotal moment, challenging the long-held belief that scientific progress might not yield a new, sustainable power source. As we await the practical realization of controlled fusion energy, a more immediate and pragmatic approach is the adoption of a diversified energy strategy. This concept can be likened to the balanced and varied diet recommended by health professionals, symbolized by the food pyramid. Similarly, an "energy pyramid" (Fig. 1) would advocate for a balanced utilization of both renewable and nonrenewable energy sources. This pyramid is expected to encompass a mix of renewable energy sources such as hydro, solar, and wind power, along with biofuels and, eventually, energy from nuclear fusion.

The global focus on developing a Hydrogen Economy is evident, with significant attention from the European Union, Japan, China, and the United States. Experts in energy economics anticipate that hydrogen energetics will play a crucial role in shaping the future of human civilization and currently form a significant part of political and educational discourse worldwide. Unlike coal and petroleum products, which are nonrenewable and extracted from the earth, hydrogen serves as an energy carrier, transmitting energy from its generation source to the point of use in various forms such as chemical, electrical, or mechanical energy.

This book is intended to serve as both a comprehensive reference and an introductory guide to hydrogen technology. Its primary audience includes the academic community, educators, students in engineering, economics, liberal arts, environmentalists, professionals in energy generation and transportation, political analysts, and anyone interested in sustainability and climate control. The secondary audience comprises professionals and political leaders focused on sustainability and the stewardship of natural resources for future generations.

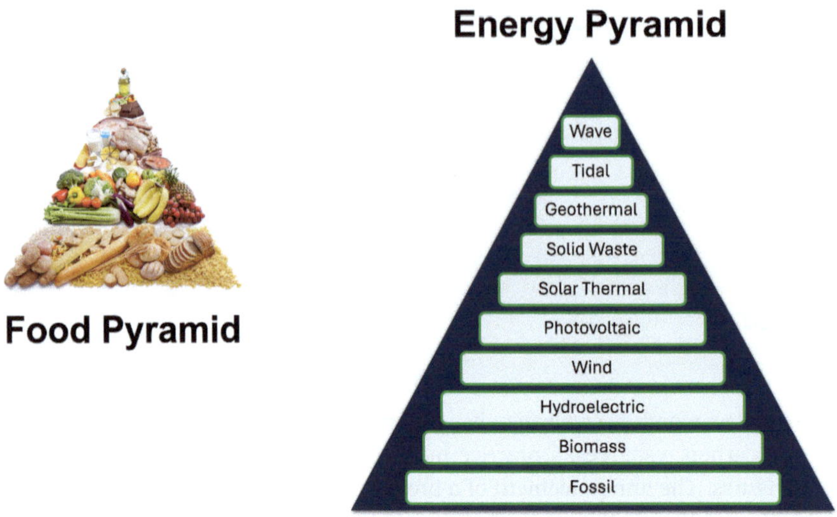

Fig. 1 Energy Pyramid

Contrary to the common perception that hydrogen energetics are limited to fuel cell transportation and power generation, this publication aims to describe all aspects of hydrogen use, from industrial applications to rocket fuels. A significant portion is dedicated to analyzing the challenges and potential benefits of integrating hydrogen into our civilization's energy sector.

This publication summarizes contemporary achievements in hydrogen technology, outlines current progress in hydrogen energetics, and addresses the global demand for new energy sources. It presents a systematic analysis of the advantages and concerns surrounding the practical use of hydrogen, emphasizing infrastructure development and the economics of hydrogen utilization. This book aims to fill the current gap in comprehensive hydrogen information, offering fundamental knowledge alongside practical aspects in a single, compact source.

1

Introduction

1.1 Terminology

Energetics: This scientific field focuses on the properties and transformation of energy in physical, chemical, or biological processes. It encompasses the study of energy flows and conversions in various systems, ranging from molecular to ecological scales. Energetics is fundamental in understanding how energy is harnessed and utilized in different contexts, from industrial processes to natural phenomena.

Energy Carrier: Hydrogen serves as an energy carrier, facilitating the transfer of energy from its generation source to the point of use in various forms, such as a chemical product, electricity, or mechanical energy. Similar to electricity, which is also an energy carrier, hydrogen does not create energy but stores and transports it. Energy carriers like hydrogen and electricity enable the convenient use of energy in forms more practical than their original sources. Hydrogen, distinct from electricity, can be stored in large quantities for future use or transported to where it is needed.

Hydrogen Economy: The hydrogen economy extends beyond traditional industrial uses of hydrogen as a feedstock. It encompasses the use of hydrogen for decarbonizing various sectors currently reliant on fossil fuels, such as transportation and steel production. This concept is integral to the low-carbon economy, aiming to minimize fossil fuel combustion and associated emissions that contribute to climate change.

Nonrenewable Energy: Major nonrenewable energy resources include coal, oil, natural gas, and nuclear energy. These fossil fuels, formed over millions of years from decomposed plants and animals, are finite and nonreplenishable. Our reliance on these resources, which we have significantly depleted over the past two centuries, poses a sustainability challenge, as they cannot be replaced once exhausted.

Hydrogen Energetics: A Future Clean Energy Carrier, First Edition. Roman J. Press.

Renewable Energy: Renewable energy sources encompass wind, solar, biomass (including sewage and ethanol), hydropower, and geothermal power. Biomass, derived from organic plant and animal materials, is a key renewable resource that replenishes through natural processes within a human timescale. Renewable resources, integral to Earth's ecosphere, are sustainable when their recovery rate matches or exceeds consumption. Life-cycle assessments are crucial in evaluating the sustainability of these resources. While oceans and agricultural outputs are often considered renewable, wood is a borderline case due to its longer regeneration time.

Sustainability: Defined by the UN World Commission on Environment and Development, sustainability is the ability to meet present needs without compromising future generations' ability to meet theirs. The UCLA Sustainability Committee further elaborates on this concept, emphasizing the integration of environmental health, social equity, and economic vitality. Sustainable practices acknowledge the finite nature of resources and advocate for their conservative and wise use, considering long-term priorities and the interconnectedness of environmental, social, and economic factors. Ultimately, sustainability is about leaving a viable world for future generations.

Decarbonization: Decarbonization refers to the process of reducing carbon dioxide emissions through the use of low-carbon power sources, which emit fewer greenhouse gases (GHGs). This term should not be confused with decarburization, which pertains to the loss of carbon in the surface layer of a material.

Global Warming: Global warming is typically used to describe the human-induced warming of the Earth's system. In contrast, climate change can encompass both natural and anthropogenic changes in the Earth's climate.

Greenhouse Gas (GHG): Greenhouse gases, or GHGs are atmospheric gases that trap heat. During daylight, the sun's rays warm the Earth's surface. At night, the Earth cools, releasing heat back into the atmosphere. However, GHGs, including carbon dioxide and chlorofluorocarbons, retain some of this heat, contributing to the greenhouse effect by absorbing infrared radiation.

Powerfuels: Powerfuels are synthetic gaseous and liquid fuels produced from green electricity. They are envisioned for use in sectors where alternatives are currently unviable, such as aviation and certain shipping methods. The development and commercialization of powerfuels are in nascent stages, with cost being a significant challenge. Currently, they are priced at €3–5/L, making them 5–10 times more expensive than fossil fuels.

Precession: Precession refers to the slow change in the direction of the axis of a rotating body, such as the Earth. This phenomenon, which can be observed in planets and stars, results in cyclic climate changes on Earth, characterized by extended periods of warming and cooling.

Energy Storage: Energy storage encompasses devices that minimize imbalances between energy demand and production. Commonly known as accumulators or batteries, these devices can store energy in various forms, including radiation, chemical, gravitational potential, electrical potential, electricity, elevated temperature, latent heat, and kinetic energy. The process involves converting energy from forms that are challenging to store into more storable and economically viable forms.

Color Code for Hydrogen: The classification of hydrogen production methods is often distinguished by their associated GHG emissions or environmental impact. In the energy sector, professionals refer to hydrogen using various color codes, each denoting the relationship between hydrogen production methods and the resultant GHG emissions.

Green Hydrogen: Green hydrogen refers to hydrogen produced through electrolysis using renewable energy sources such as wind, water, or solar power. This process emits no GHGs, contributing to a reduction in our carbon footprint. The purity of hydrogen, crucial for fuel cell applications, is often described as "five nines" (99.999%). The cost of producing green hydrogen ranges approximately from $2.5 to 6.0/kg.

Gray Hydrogen: Gray hydrogen is produced from natural gas via steam-methane reforming, similar to blue hydrogen, but without any measures to capture the carbon dioxide by-products. The cost of gray hydrogen production is around $1.5/kg, varying with natural gas prices and carbon emission factors.

Blue Hydrogen: Blue hydrogen is generated from natural gas through steam methane reforming, where natural gas reacts with steam and a catalyst to produce hydrogen and carbon monoxide. Further reaction with water yields more hydrogen and carbon dioxide. When the CO_2 emissions are captured and stored, the process is considered carbon-neutral, and the resulting hydrogen is termed "blue hydrogen." Blue hydrogen production costs about $2.5/kg, positioning it between gray and green hydrogen in terms of expense. The feasibility of converting existing hydrogen production facilities to blue hydrogen production depends on the availability of CO_2 storage capacity.

However, blue hydrogen production is subject to controversy due to methane emissions associated with natural gas extraction and transportation. Methane, though shorter-lived in the atmosphere than CO_2, is a potent GHG, with one ton of methane equating to 28–36 tons of CO_2 in terms of global warming potential.

Pink Hydrogen: Pink hydrogen is produced via electrolysis powered by nuclear energy, which does not emit carbon dioxide. However, the process generates radioactive waste, necessitating safe storage for thousands of years.

Yellow Hydrogen: Yellow hydrogen is produced through electrolysis using energy from the electrical grid. The carbon emissions associated with yellow hydrogen vary significantly based on the energy sources powering the grid.

Turquoise Hydrogen: Turquoise hydrogen refers to hydrogen generated from methane pyrolysis, a process that splits methane into hydrogen and solid carbon using heat in reactors. The solid carbon by-product can be sold separately, adding economic value. While still in the early stages of commercialization, turquoise hydrogen has the potential to be cost-efficient and environmentally conscious, especially when powered by clean energy sources.

Solid Hydrogen: Solid hydrogen typically refers to hydrogen in its solid state, achieved at temperatures below its melting point of 14.01 K (−259.14 °C). Another application of the term "solid hydrogen" pertains to metal hydride storage devices, which use a solid-state medium for hydrogen storage.

1.2 Sustainability and Climate Change

Sustainability, a concept deeply rooted in the understanding that human activities are a primary driver of recent climatic changes, emphasizes the need for responsible stewardship of the environment. The escalation of GHG emissions, particularly carbon dioxide, has led to a significant increase in global temperatures, a phenomenon faster than any previously recorded. Scientific consensus indicates that anthropogenic factors, especially since the Industrial Revolution in the 1800s, are the primary contributors to these emissions. The burning of fossil fuels, releasing heat-trapping gases, is largely responsible for the observed climate change, encompassing everything from global warming to habitat destruction. This change is not just a scientific observation but is becoming increasingly evident in its impact on human lifestyles and the global ecosystem.

The interaction between humans and nature has historically been complex. Human activities, such as mining, deforestation for agriculture, and the introduction of invasive species, have significantly altered landscapes and ecosystems. Conversely, natural phenomena like volcanic eruptions, floods, and earthquakes have also shaped human history. The rapid technological advancement and global expansion of human populations have led to profound changes in the natural world, including the atmosphere. It is a sobering realization that resources formed over millions of years have been substantially depleted in just two centuries. Human-induced changes, whether through land clearing for agriculture or industrial pollution from urbanization, have led to a decrease in biodiversity and ecosystem stability.

1.3 Decarbonization

The philosophy of sustainability urges us to avoid burdening future generations with the ecological, human, and economic consequences of our actions. Decarbonization encompasses a range of measures aimed at significantly reducing and offsetting carbon dioxide emissions. This process is crucial in mitigating climate change and involves various strategies and technologies.

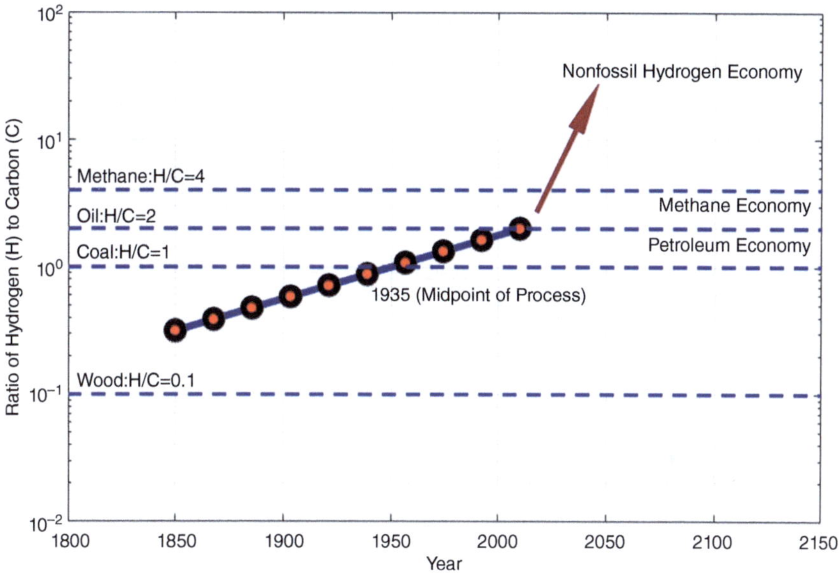

Fig. 1.1 Ratio of Hydrogen to Carbon Energy Consumption Since 1860.

Figure 1.1 illustrates the global shift in primary energy consumption from wood to fossil fuels and, more recently, toward a hydrogen economy. This transition, known as decarbonization, is evident in the United States, as outlined in the DOE Industrial Decarbonization Roadmap. This roadmap targets five major CO_2-emitting industries: petroleum refining, chemicals, iron and steel, cement, and food and beverage. These sectors account for approximately 51% of the energy-related CO_2 emissions in the US industrial sector and 15% of the nation's total energy-related CO_2 emissions, equating to about 1,360 million tons of CO_2 in 2020. The DOE's commitment to decarbonizing the industrial sector aims to create sustainable jobs, spur economic growth, and foster a cleaner, more equitable future for all Americans.

1.4 Climate Change

A critical piece of evidence for anthropogenic climate change is the Keeling Curve, a dataset that has tracked atmospheric carbon dioxide levels since 1958. Named after Charles Keeling, the researcher who initiated these measurements, this curve reveals a significant and exponential increase in atmospheric CO_2 over time. Carbon dioxide is produced both naturally, as in animal respiration and methane emissions from livestock, and anthropogenically, through activities like electricity production and transportation. The Keeling Curve (Fig. 1.2) graphically

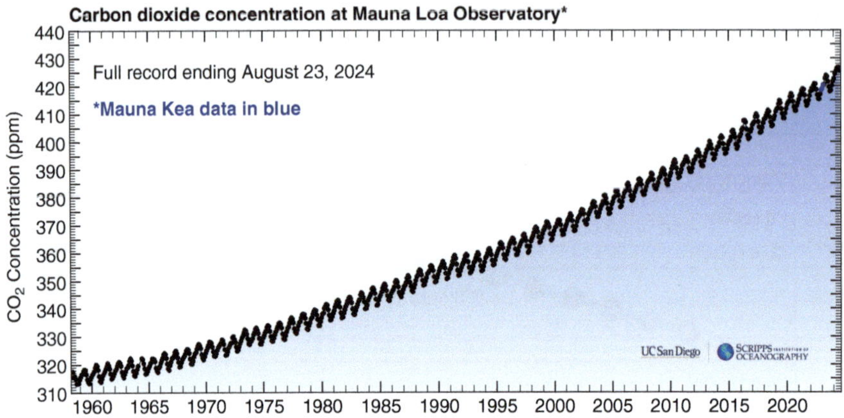

Fig. 1.2 Keeling Curve Showing Carbon Dioxide Concentration at Mauna Loa Observatory. *Sources:* Regents of the University of California/ https://bluemoon.ucsd.edu/co2_400/mlo_full_record.pdf, last accessed on March 28, 2024

illustrates this rise in CO_2 concentration, measured in parts per million (ppm). Human activities, including agriculture and the burning of fossil fuels since the Industrial Revolution, have been primary contributors to this increase.

Climate change is an established reality, with long-term impacts such as ice melt, ocean warming, sea-level rise, and altered marine ecosystems. The consensus is that the magnitude of climate change is largely driven by anthropogenic CO_2 emissions, which have led to a significant increase in Earth's temperatures. However, attributing climate change solely to human activities necessitates further evidence. The role of human actions in atmospheric changes and their impact on solar activities is complex and requires additional research. For instance, the process of photosynthesis in green vegetation, converting sunlight, carbon dioxide, and water into carbohydrates is a fundamental biological process.

Anthropogenic climate change, also referred to as habitat destruction or global warming, signifies a change in climate primarily caused by human activities. We are now entering a new geological era termed the "Anthropocene," characterized by a climate markedly different from that known to our ancestors and evidenced by rising global temperatures.

Human influence on nature has been profound for millennia. Activities such as mining, agriculture, and the introduction of invasive species have significantly altered natural environments and ecosystems. These anthropogenic changes, ranging from deforestation to urbanization, have led to decreased biodiversity and ecological imbalances.

The increase in GHGs and the resultant global climate alteration are prime examples of anthropogenic change, a phenomenon that has become increasingly apparent over recent decades. The complexity of Earth's systems and the challenge of distinguishing natural variations from human-induced changes contribute to the difficulty in fully understanding anthropogenic climate change.

Interestingly, elevated CO_2 levels can have both negative and positive effects on the environment. For example, some models suggest that increased temperatures could lead to higher humidity, more vegetation, reduced ice cover, and rising sea levels. Climate science encompasses the study of air and water movement, chemical and biological processes, and various other factors, making it a challenging field due to the interplay of numerous complex processes. Even contemporary phenomena like the increased electricity consumption for Bitcoin mining could have a significant environmental impact.

There is no unanimous scientific consensus on the negative impacts of increased CO_2 levels. Some theories, primarily based on geological evidence, propose that rising temperatures could enhance humidity and vegetation growth, leading to reduced ice cover and higher sea levels. The extent to which temperatures might rise in the next millennium remains uncertain, but the influence of Earth's precession on climate change cannot be ignored.

Regarding CO_2 emissions, it is important to note that CO_2 is a vital component for plant life and, by extension, all biological life on Earth. While plants absorb CO_2 and humans exhale it, a significant portion of biomass does not decompose and is sequestered in soil and ocean sediments. Thus, the introduction of new CO_2 into the atmosphere is essential for the continuation of the biosphere. Volcanic eruptions also contribute CO_2 to the atmosphere, but contrary to some beliefs, the total CO_2 emissions from volcanic activity are only a fraction of those produced by human activities.

While the role of anthropogenic activities in climate change is significant, it is part of a complex interplay of natural and human-induced factors. The challenge lies in understanding these interactions and their implications for the future of our planet (Fig. 1.3).

The CO_2 emissions from human activities are substantial. In 2021, global CO_2 emissions from energy combustion and industrial processes alone reached a record high of 36.3 gigatons, as depicted in Fig. 1.4.

We should not ignore significant emisions of the GHGs in forms of methane, contributed by livestock and related manure management. In accordance to Food and Agriculture Organization of the UN estimation of the livestock produces 16.1% of the global GHGs emissions. The other studies conclude that such emissions can reach up to 19.6%.

The climate system is incredibly complex, encompassing the movement of water and air masses, the influences of chemistry and biology, and processes like vaporization, ice melting, and the reflection and absorption of sunlight. This

Fig. 1.3 Volcano Eruption. *Sources:* Jeronimo Ramos/Adobe Stock Photos

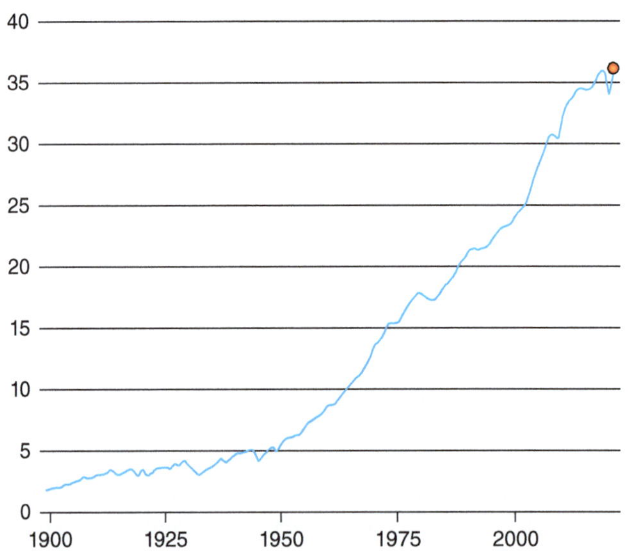

Fig. 1.4 CO_2 Emissions from Energy Combustion and Industrial Processes 1900–2021. *Sources:* International Energy Agency/https://www.iea.org/data-and-statistics/charts/co2-emissions-from-energy-combustion-and-industrial-processes-1900-2021, last accessed 26 March 2024/CC BY 4.0

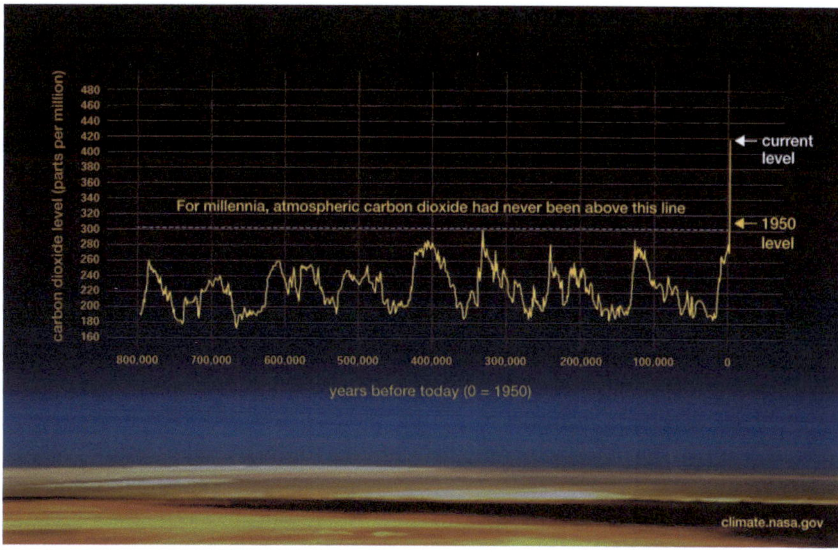

Fig. 1.5 The Relentless Rise of Carbon Dioxide. *Sources:* NASA Climate/The National Aeronautics and Space Administration/Public Domain

complexity makes it challenging to distinguish between natural variations and changes caused by human activities (Fig. 1.5).

While the correlation between fossil fuel emissions and planetary temperature is often attributed to GHG emissions, other factors, such as astronomical events, also play a significant role in global warming. Over 2,000 years ago, ancient Egyptian astronomers discovered the phenomenon of Earth's precession – a slow, cyclic movement of the Earth's axis relative to the Sun. This gradual shift, occurring over approximately 25,772 years, results in different parts of the planet receiving varying amounts of sunlight over millennia, thereby influencing the global climate.

Human lifespans are too short to observe the direct effects of precession on climate change. However, it is evident that Earth's climate – encompassing temperature, humidity, ocean levels, and ecological systems – is influenced by this astronomical cycle. Precession, a lesser-known motion of the Earth alongside its daily rotation and annual revolution, results in significant long-term climate variations.

The last Ice Age in Europe ended approximately 12,000 years ago. As the Earth's axis continues its precessional movement, the Northern Hemisphere has yet to experience the peak of this warming cycle, which is expected to occur in about a thousand years. Following this peak, a gradual cooling will lead to the onset of another Ice Age in approximately 13,000 years. During this "peak of cold," regions like Scandinavia, northern Europe, Siberia, and Canada will experience significant cooling, and ocean levels will drop considerably. Interestingly, Antarctica is predicted to become ice-free and potentially habitable during this period.

1.5 Energy Ethics

Human civilization's quest for energy has profoundly impacted our planet. In just a few generations, we are on the brink of depleting fossil fuel reserves that took hundreds of millions of years to form. The combustion of coal and oil releases sulfur dioxide (SO_2) into the atmosphere at rates surpassing natural sources, such as the decomposition of dimethyl sulfide from oceans. Similarly, the emission of nitrogen oxides (NO_x) from burning fossil fuels and biomass now likely exceeds natural emissions, contributing to photochemical smog in urban areas. GHGs like carbon dioxide (CO_2) and methane (CH_4), often linked to energy production, are accelerating the global warming process. These impacts have led to discussions about defining our current geological period as the "Anthropocene Epoch."

A comprehensive energy policy is essential to promote energy conservation, minimize environmental impact, and facilitate a transition to sustainable environmental management at all societal levels – from individuals and families to local, national, and international communities. As consumers and society members, we must understand the connection between our consumption choices and global impacts. Adopting energy-efficient appliances, homes, and vehicles, especially for long-distance single-person travel, is crucial. Opting for hybrid vehicles, locally sourced products, installing solar panels, and purchasing green power are steps toward meeting future energy needs while preserving the environment. Additionally, recycling and choosing recycled materials over new products contribute to energy conservation.

Government policies play a pivotal role in promoting energy efficiency and environmental improvement. Taxes, subsidies, incentives, and standards should encourage energy-efficient practices in automobiles, factories, appliances, and homes. In some countries, robust bike infrastructure, public transportation, and high vehicle registration taxes have shifted transportation preferences from cars to bikes, subways, or buses. Notably, countries with higher energy prices often have lower per capita energy needs compared to those with lower energy prices.

To ensure the sustainability of human civilization, society must focus on improving energy efficiency and embracing alternative energy sources, including hydrogen technology. Transitioning to renewable energy sources and hydrogen technology is a significant step toward sustainable environmental management.

Table 1.1 Anthropological Factors vs. Nonanthropological Factors

Anthropological Factors	Nonanthropological Factors
• Transport, using internal combustion	• Precision Earth movement
• Fossil fuels mining	• Sun's activities
• Agricultural activities	• Other planets and asteroids influence
• Military actions	• Volcano eruptions
• Crypto currency mining	• Flora photosynthesis
• Fertilizers production	• Fauna emissions
• Chemical products	
• Energy production	
Global Warming Verdict	
• Impossible to reject the human influence on our planet future	
• Impossible to ignore the nonanthropological factors on our planet future	

An additional aspect of sustainability concerns the aftermath of military operations, which often leave behind minefields, destroyed infrastructure, and air contamination. While not the primary focus of this publication, the absence of war is crucial for preserving our planet for future generations, arguably as important as addressing potential climate change. The cessation of conflict and the promotion of peace are integral to the long-term sustainability of our environment and civilization.

The main factors influencing climate change are presented in Table 1.1.

1.6 Hydrogen Economy: Pros and Cons

1.6.1 Incentives for Transition to Hydrogen

The quest for energy has evolved significantly over time. Our ancestors harnessed natural forces like water, animals, and wind. This was followed by the use of coal, oil, and gas. However, in the past two decades, concerns over pollution, habitat destruction, the finite nature of fossil fuel resources, and their impact on climate change have spurred the search for alternative energy sources, with hydrogen emerging as a promising option.

Key incentives for transitioning to a hydrogen-based society include the following:

- Hydrogen as an ideal alternative to diminishing natural resources.
- Mitigating global warming and climate change.
- Improving air quality.

- Protecting and conserving environmental resources and landmarks.
- Promoting energy decentralization.
- Reducing dependence on energy imports.

The exponential increase in the global population, coupled with rising food production and energy demands, underscores the need for alternative energy sources. Hydrogen, along with electricity, offers pollution-free energy solutions. Electricity, however, faces challenges in delivery, especially to remote areas, requiring extensive infrastructure and posing storage difficulties. The transition to zero-emission vehicles (ZEVs) has been bolstered by advancements in fuel cell technology. Although hydrogen storage and transportation present challenges, it holds potential for a wireless society.

The European Union's commitment to hydrogen energy is evident in its 2020 strategy, "A hydrogen strategy for a climate-neutral Europe." This strategy aims to develop a comprehensive hydrogen ecosystem, from research to production scale-up and infrastructure development. The strategy recognizes hydrogen's versatility as a feedstock, energy carrier, and storage medium, capable of spawning new products across various sectors. Crucially, hydrogen's near-zero emissions make it a key player in decarbonizing industries where reducing carbon emissions is both urgent and challenging, such as in the chemical and metallurgical sectors. This aligns with the EU's goal of achieving carbon neutrality by 2050 and supports global efforts under the Paris Agreement.

Japan, too, is actively pursuing hydrogen energy. The government has revised its plans to increase annual hydrogen production sixfold to 12 million tons by 2040, with a commitment of $107 billion over the next 15 years to develop hydrogen supply chains. This investment aims to replace fossil fuels in carbon-intensive industries, enhancing global competitiveness in sectors like steel and chemicals.

1.6.2 Existing Limitations

Many critics point out several challenges in transitioning to a hydrogen economy, contrasting it with the current reliance on conventional fossil fuels. The primary critique centers on the absence of naturally occurring hydrogen sources, unlike fossil fuels which can be extracted directly from the Earth.

Hydrogen's highly reactive and lightweight nature causes it to escape easily into the atmosphere, making direct extraction impractical. However, extracting hydrogen from compounds like water or metal hydrates is feasible. Contrary to popular belief, there are instances of naturally occurring hydrogen, such as in Bourakebougou, Mali, where it is used for electricity generation. While the discovery of more natural hydrogen sources is possible, relying on them as a primary energy solution remains uncertain.

Critics also highlight the lack of efficient and long-term storage solutions, the current high cost of hydrogen production, underdeveloped infrastructure, and substantial financial requirements for technology and infrastructure development. Overcoming these challenges will require concerted efforts from scientists, government involvement, and social advocacy.

Patrick Moore, a co-founder of Greenpeace and environmental scientist, offers a perspective on the environmental movement. He observes that the younger generation is often led to believe that human activities are significantly harming the Earth, leading to feelings of guilt. Moore points out that some activist claims lack scientific grounding. For instance, contrary to the narrative of polar bear extinction due to melting Arctic ice, their population has actually increased from about 6,000 to 8,000 in 1973 to an estimated 26,000 today.

There is also debate over the role of CO_2 in global warming. The term "carbon neutrality" is more political than scientific. CO_2, comprising carbon and oxygen, is essential for plant growth and is fundamental to all life on Earth.

Regarding renewable energy sources like solar and wind, critics argue that they are costly and unreliable. These energy sources require vast land areas, are subject to weather conditions, and often need backup from nonrenewable energy sources.

The introduction of hydrogen as a new energy carrier, coupled with advancements in fuel cell technology, marks a significant step toward a hydrogen economy. Government support and the availability of low-cost, clean electricity for fuel cell transportation are crucial for kickstarting the hydrogen economy.

2

Energy Resources

2.1 Nonrenewable Energy Sources

Nonrenewable energy sources are those that are depleted within a foreseeable timeframe, typically a few hundred years. These sources currently dominate our energy consumption. An interesting case is firewood, often considered nonrenewable due to the lengthy time required for its restoration. The most common nonrenewable sources are fossil fuels, formed over 300 million years ago during the Carboniferous Period. This era was marked by swampy forests transforming into coal and wetland microorganisms giving rise to petroleum and natural gas. Long before the age of dinosaurs, the Earth's landscape was diverse, and dead plant matter was compressed under the seabed. Over time, layers of rock and sediment created high heat and pressure, converting these organic remains into fossil fuels. Today, vast underground reservoirs of these nonrenewable energy sources exist across various continents.

The primary concern with nonrenewable energy sources, particularly fossil fuels, is their environmental impact. Despite efforts to remove toxic by-products like nitrogen and sulfur oxides, fossil fuels remain significant contributors to greenhouse gas emissions.

Another nonrenewable energy source is nuclear fuel, derived from the fission of heavy nuclei such as Uranium-235. Additionally, oil sands and natural gas hydrates exist as potential energy sources, but their commercial use is limited due to the high costs of extraction and production.

2.1.1 Advantages and Disadvantages

Fossil fuels are a valuable and reliable energy source, adaptable to various climates and times. They are relatively inexpensive to extract, store, and distribute. However, their combustion releases numerous toxic gases and pollutants into the

Hydrogen Energetics: A Future Clean Energy Carrier, First Edition. Roman J. Press.

air, water, and soil. Mining activities can lead to ground subsidence and potentially cause long-burning underground fires. Furthermore, mining poses significant risks, including exposure to toxic gases and the danger of explosions, making it one of the most hazardous occupations globally.

2.1.2 Coal

Coal has been a dependable energy source for generations, primarily used for heating and electricity generation. It is the most abundant fossil fuel, contributing to approximately 60% of global electricity production. Chemically, coal is primarily composed of carbon, along with ash (consisting of silicates and metals), and sulfur. Certain constituents in coal, particularly sulfur, are undesirable as they contribute to the formation of hazardous sulfur oxides during combustion.

Coal is found in various grades, each with distinct characteristics. Anthracite, formed under high-pressure and low-moisture conditions, boasts the highest carbon content at about 95%. Sub-bituminous coal, with a lower carbon content of around 40%, is valued for its minimal sulfur content. Lignite, formed under lower pressure and higher moisture, contains only about 25% carbon. Traditional coal mining has shifted from underground to surface mining, a safer yet more environmentally impactful method (Fig. 2.1).

Fig. 2.1 Sedimentary Combustible Rock, Coalburn Pit, Nova Scotia. *Source:* Michael C. Rygel/Wikimedia Commons/CC BY SA 3.0

Table 2.1 World Coal Recoverable Reserves

- United States 22%
- Russia 15%
- Australia 14%
- China 14%
- India 11%

Source: US Energy Information Administration https://www.eia.gov/energyexplained/coal/how-much-coal-is-left.php/Public Domain

As of October 2022, the estimated global recoverable coal reserves were about 1,161 billion short tons, with the top 5 countries holding approximately 75% of the world's proven coal reserves (Table 2.1).

Typically, up to 90% of coal is used within its country of origin due to transportation challenges compared to oil and natural gas. The combustion of coal necessitates the use of scrubbers or filters to mitigate the release of environmentally harmful gases. Coal reserves are projected to last around 200 years, the longest among nonrenewable fuels. In the United States, based on 2021 production rates, coal reserves could last approximately 435 years.

Coal is a black or brownish-black combustible rock primarily composed of carbon, with varying amounts of hydrogen, sulfur, oxygen, and nitrogen. When burned, coal yields valuable by-products used in the production of chemicals, plastics, cement, and building materials. The quality of coal is determined by its level of carbonization, with peat being the lowest grade and anthracite, predominantly found in mountainous regions, being the highest. Anthracite, characterized by high pressure and low moisture during formation, has the highest carbon content. Sub-bituminous coal, with a carbon content of about 40%, is notable for its low sulfur content. Lignite, formed under lower pressure and higher moisture, contains only 25% carbon. The shift from underground to surface mining has reduced human casualties but increased environmental impact.

Coal can be converted into "coal gas" (used as a substitute for natural gas) and syngas (a mixture of carbon monoxide and hydrogen). As oil and natural gas prices rise, these alternatives become more attractive. Coal is also liquefied through the Fischer–Tropsch process to substitute for oil. In the steel industry, coal is vital for producing pig iron and steel and as a reinforcing agent in the plastics and rubber industry.

2.1.3 Petroleum

Petroleum, formed over 100 million years ago, originated from microorganisms buried under layers of sand and silt. Subjected to intense heat and pressure beneath the Earth's surface, these organic materials transformed into waxy solids

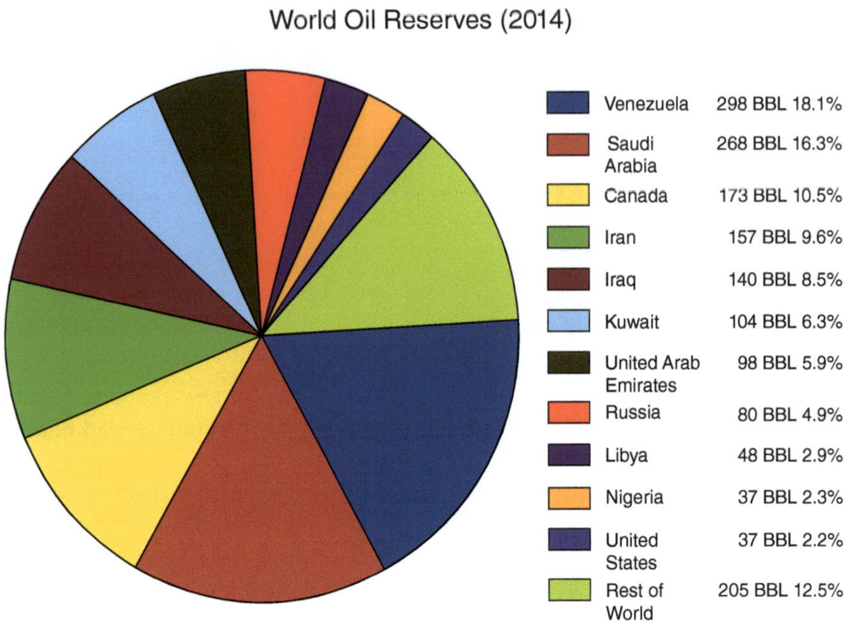

Fig. 2.2 World Oil Reserves by Country. *Source:* Delphi234/Wikimedia Commons/ CC BY 1.0

known as kerogen, eventually evolving into gaseous, liquid, and solid hydrocarbons. Crude oil, a yellowish-black viscous liquid, is a complex mixture of these hydrocarbons. In refineries, crude oil undergoes distillation, separating it into purer fractions. Figure 2.2 outlines the regions and major countries with significant oil reserves.

Oil extraction typically involves drilling into subterranean reservoirs or from ocean platforms (Fig. 2.3).

In 2004, the world's total oil reserves were estimated at 148 gigatons (Gt). The Middle East is a major oil-rich region, holding about 65% of the world's reserves. Key oil-producing countries include Saudi Arabia, Iran, Iraq, Kuwait, and the United Arab Emirates (Fig. 2.4). Other significant non-OPEC producers are Russia, Venezuela, and Mexico. Canada also has substantial reserves, primarily in oil sands.

Estimates vary on when world oil production will peak, with predictions ranging from 1980 to 2020. M. King Hubbert's 1956 prediction, often criticized yet still relevant, suggested a peak within this timeframe (Fig. 2.5). At the current consumption rate, petroleum reserves could be depleted in the next 40–50 years. The United States, despite being a major producer, imports over half of its petroleum needs, increasingly supplying Europe due to recent political dynamics.

Fig. 2.3 Oil Ocean Platform. *Source:* With permission of Siemens Energy

Most petroleum is refined into gasoline, with over half used for various fuels, including diesel, heating oil, and aviation fuel. Beyond fuel, petroleum is a crucial raw material for manufacturing plastics, rubber, and pharmaceuticals.

2.1.4 Natural Gas

Natural gas, like oil, is formed through anaerobic processes involving the decay of microorganisms over millions of years. It is utilized for heating, cooking, electricity generation, and in the production of various chemicals. Natural gas can be converted into liquid natural gas (LNG), a cleaner and more efficient form for distribution, as it occupies significantly less space than its gaseous counterpart (see Fig. 2.6). For example, the volume of natural gas that fills a large beach ball in its gaseous state can be compressed into the size of a ping-pong ball when liquefied. LNG's ease of storage and versatility make it an attractive alternative to gasoline in many countries.

Natural gas deposits are often trapped in extensive rock formations. Extraction methods include hydraulic fracturing, or fracking, which involves using high-pressure water to fracture the rocks and release the trapped gas. If the rock is particularly hard, acid may be used to dissolve it, or small grains of glass or sand can be employed to keep the fractures open and facilitate gas escape.

2.1.4.1 Advantages and Disadvantages

Natural gas is a relatively inexpensive and cleaner fossil fuel compared to oil or coal. Its combustion primarily releases carbon dioxide and water vapor. However, its extraction poses environmental challenges. Fracking can induce minor seismic

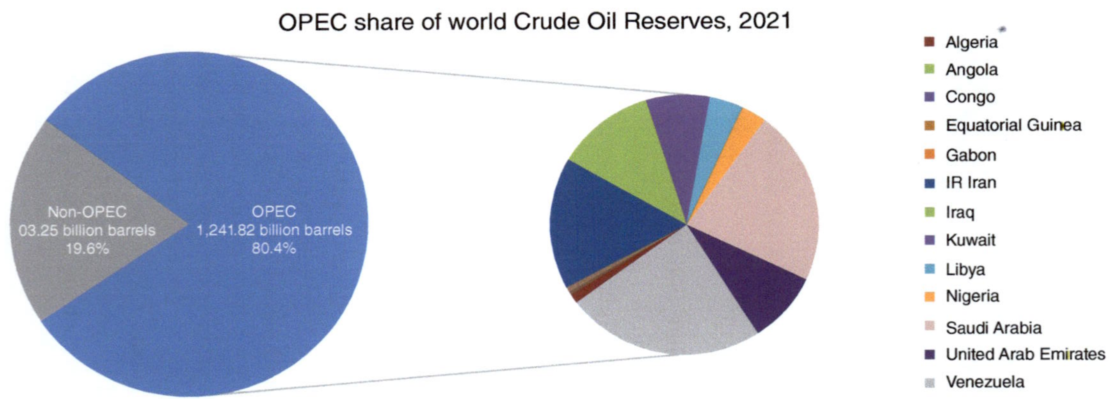

OPEC proven crude oil reserves, at end 2021 (billion barrels, OPEC share)

Venezuela	303.47	24.4%	United Arab Emirates	111.00	8.9%	Algeria	12.20	1.0%	Equatorial Guinea	1.10	0.1%
Saudi Arabia	267.19	21.5%	Kuwait	101.50	8.2%	Angola	2.52	0.2%			
IR Iran	208.60	16.8%	Libya	48.36	3.9%	Gabon	2.00	0.2%			
Iraq	145.02	11.7%	Nigeria	37.05	3.0%	Congo	1.81	0.1%			

Fig. 2.4 OPEC Share of World Crude Oil Reserves, 2021. *Source:* OPEC Annual Statistical Bulletin 2022/Organization of the Petroleum Exporting Countries

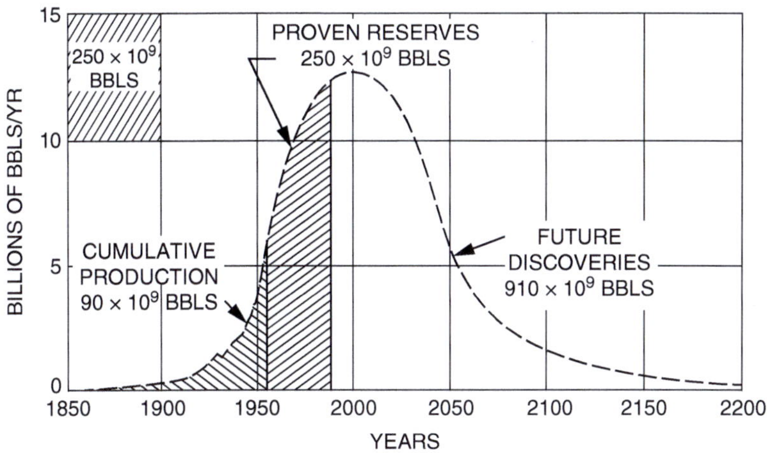

Fig. 2.5 The Hubbert Curve. *Source:* Hankwang/Wikipedia Commons/CC BY-SA 3.0

Fig. 2.6 Natural Gas Shipment. *Source:* AA+W/Adobe Stock Photos

activities and potentially contaminate water sources with high-pressure water and chemicals used in the process.

Natural Gas Production by Countries (billion cubic meters) (Table 2.2).

Natural gas primarily consists of methane (CH_4), with concentrations ranging from 50% to 90%, and the remainder being other hydrocarbons with 2–4 carbon

Table 2.2 Natural Gas Production by Countries

United States: 934
Russia: 701
Iran: 257
China: 209
Canada: 172
Qatar: 177
Australia: 117
Norway: 114

Source: Adapted from https://investingnews.com/top-natural-gas-producers/

atoms. Table 2.1 lists countries with the largest natural gas reserves. It is often transported via pipelines or as LNG at low temperatures (−160 °C) in containers. LNG's lower viscosity makes it easier to transport and process than oil. It is widely used as a fuel in industry and residential settings and as a precursor for plastics and pharmaceuticals production.

Natural gas, being a purer hydrocarbon mix, produces fewer environmentally harmful by-products when burned compared to oil. For instance, emissions of nitrogen oxides or particulate matter are minimal. An additional advantage is its cost-effectiveness relative to oil. However, natural gas is highly flammable and potentially hazardous. For safety, sulfur-containing compounds, known as mercaptans, are added to natural gas to make leaks detectable by their distinct garlic-like odor.

The world's natural gas reserves are projected to last about 60 years, slightly longer than oil reserves.

2.2 Renewable Energy

Renewable energy, often referred to as clean, green, or alternative energy, is an environmentally friendly substitute for conventional fossil fuels. It is considered an inexhaustible resource over time, although its availability can be intermittent (Fig. 2.7).

In France, renewable energy constituted 27% of the total electricity consumption in 2020 (source: RTE). On February 10, 2022, the French government announced plans to double electricity production from renewable sources by 2030, allocating 1 billion euros for innovation in this sector.

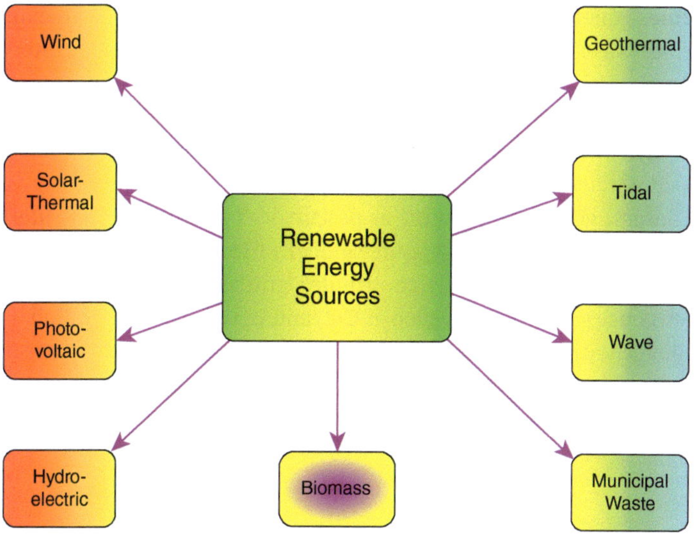

Fig. 2.7 Renewable Energy Sources

In the United States (Fig. 2.8), renewable energy accounts for about 20% of the total electricity generation, a figure that is steadily increasing. The following graphic illustrates the distribution of electricity production in 2021 among various types of renewable energy.

The primary source of renewable energy is the sun, either directly or indirectly influencing Earth's internal and external dynamics. Solar heat and terrestrial

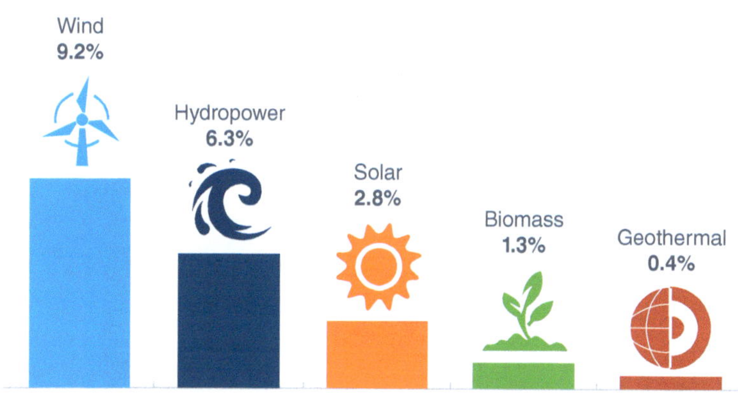

Fig. 2.8 Renewable Energy Production in the United States. *Source:* US Department of Energy/https://www.energy.gov/eere/renewable-energy, last accessed 26 March 2024/ Public Domain

surface temperatures generate winds, which, along with tidal forces, create ocean currents and surface waves. These winds, combined with solar heat, drive water evaporation and precipitation, resulting in flowing rivers, lakes, and streams. Sunlight and water facilitate the growth of vegetation, which can be harnessed as organic matter for energy. These natural phenomena can be converted into renewable sources of heat, electricity, or fuel.

While oil and natural gas currently serve as precursors for hydrogen as a fuel carrier, coal is the fastest-growing energy source globally. However, all nonrenewable energy sources, including fossil fuels, contribute significantly to environmental harm and greenhouse gas emissions, such as CO_2 and CH_4, exacerbating global warming. A challenge with hydrogen as a renewable energy source is that most current hydrogen production methods rely on nonrenewable resources, such as synthesizing gas from coal or converting natural gas.

2.2.1 Solar Energy

Solar energy stands as the most effective and stable renewable energy source on Earth (Fig. 2.9). It is a sustainable and nonpolluting energy source. However, the current technology, primarily based on solar cells with relatively lower efficiency, requires substantial land occupation, often in remote desert areas far from major human settlements. The ecological impact of such extensive land use warrants further investigation.

Fig. 2.9 The Troia photovoltaic solar farm in Apulia, Italy. *Source:* With Permission of Siemens Energy

The sun emits radiation at a temperature of approximately 6,000°K, spanning ultraviolet, visible, and near-infrared wavelengths. The Earth's ozone layer absorbs ultraviolet photons, allowing only visible radiation (ranging from 4,000 to 8,000 Å) to reach the surface.

Solar radiation is converted into electricity through the photoelectric effect, utilized in photovoltaic and photoelectrochemical solar cells. Photovoltaic cells are solid-state devices that directly convert sunlight into electricity using two different semiconductors. In contrast, photoelectrochemical cells combine a semiconductor with a liquid electrolyte.

Thermal solar systems absorb the sun's radiation and convert the stored heat for various applications, including hot water preparation, building heating and cooling, and electricity generation.

The solar radiation budget is the balance between absorbed solar energy flux and outgoing long-wavelength radiation at the Earth's atmosphere's top. This balance is crucial for maintaining average climatic conditions, though it can be disrupted on smaller scales, influencing atmospheric and oceanic circulations. Climate perturbations, such as changes in vegetation or ice formation, can increase greenhouse gases like CO_2, leading to adverse effects. Understanding the interactions between clouds and radiation is vital for climate research.

In terms of energy potential, the net yearly solar energy input to the planet is immense, dwarfing current global oil consumption. Even with a conservative energy transformation efficiency of 10% and using only 1% of land surface for solar energy collectors, solar energy could provide about six times the energy equivalent of current oil usage.

2.2.1.1 Solar Energy Benefits

- Clean energy source with minimal environmental intrusion.
- Potential to provide significantly more energy than current oil usage, even with low efficiency and minimal land coverage.
- Suitable for both community and individual use, with the possibility of selling excess power to electric companies.
- Relative simplicity in connecting to the electric grid.

2.2.1.2 Solar Energy Disadvantages

- Limited availability due to the diurnal cycle of the sun.
- Lack of effective energy storage technology.
- Inadequate efficiency in energy conversion.
- Limited environmental impact mitigation.

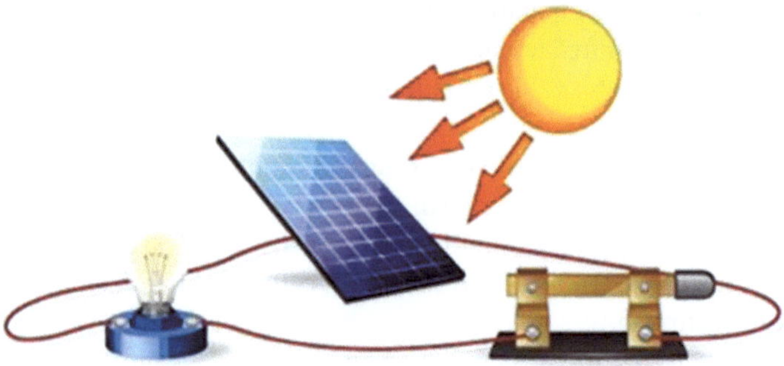

Fig. 2.10 Solar Energy Diagram. *Source:* Brgfx/Freepik Company S.L

2.2.1.3 Solar Cells

Solar cells operate on the photovoltaic effect, first discovered by Alexandre Edmond Becquerel in 1839. The first solar cell, made of selenium and gold in 1883, had an efficiency of 1%. Modern solar panels can achieve efficiencies up to 30%. However, they cannot absorb energy from the entire solar spectrum, limiting their efficiency (Fig. 2.10).

Solar panels consist of arrays of solar cells connected together. Other types of solar cells, such as photoelectrolytic cells and cells based on photoelectrosynthesis, utilize different chemical phenomena.

2.2.2 Wind Energy

Wind energy harnesses the power of atmospheric winds caused by the uneven heating and cooling of the Earth's surface. This natural phenomenon, where land surfaces heat faster than water bodies, creates a continuous cycle of air movement. Wind energy can be utilized for various purposes, including grinding grain, pumping water, and, most notably, generating electricity through wind turbines.

Wind turbines operate by capturing the kinetic energy of wind to rotate aerodynamic blades. As wind flows over these blades, a lift force is generated, turning the rotor. This rotor is connected to a drive shaft that spins a generator, converting mechanical energy into electricity. Wind turbines are effective at wind speeds of 8–16 miles/hour but shut down at speeds exceeding 65 miles/hour to prevent damage.

One wind turbine can generate between 5,000 and 5 million kWh of electricity annually. Wind farms, which are clusters of turbines, are connected to utility power grids to supply electricity on a larger scale (Fig. 2.11). Wind power can also be integrated with solar energy systems and hydrogen storage to store excess energy, producing pure water as a by-product.

Fig. 2.11 Land Wind Farm. *Source:* US Department of Energy/Public Domain

Offshore wind farms, located near or far from shorelines, are becoming increasingly popular. These installations minimize land use concerns and are connected to onshore grids via undersea cables (Fig. 2.12). Europe, particularly Germany, is advancing in this field with the development of large-scale turbines.

In 2022, global wind power electricity generation grew by 14%, reaching over 2,100 TWh. In the United States, wind energy accounts for about 10.2% of total electricity production. However, wind turbine efficiency typically ranges from 20% to 40% of nominal power due to variable wind conditions.

Fig. 2.12 Offshore Wind Farm. *Source:* With Permission of Siemens Energy

2.2.2.1 Advantages of Wind Power

- Clean and renewable energy source.
- Can be installed on land or offshore.
- Minimal environmental intrusion.
- Limited land usage.
- Suitable for remote locations and individual use.
- Excess power can be sold to electric companies.
- Simple integration with the electric grid.

2.2.2.2 Disadvantages of Wind Power

- Power output is dependent on wind availability.
- Logistical challenges and higher costs.
- Potential negative impact on bird populations.
- Complexity in dismantling previous installations.
- Lack of effective energy storage technology.
- Impact on scenic views.

Innovations in wind turbine design are ongoing, including unique blade shapes and bladeless designs. Aeromine Co. is developing a noiseless, esthetically pleasing rooftop installation with a wide operational wind speed range, potentially more efficient than solar. Blueenergy AG is working on integrated solar vertical axis wind turbines, offering quiet, space-saving options for both urban and rural settings.

2.2.3 Hydroelectric Power

Hydroelectric power, also known as hydroelectricity, stands as one of the oldest and most efficient sources of clean energy. For millennia, humans have harnessed the gravitational force of falling or fast-flowing water, initially converting it into mechanical energy and later into electricity. By the early 21st century, hydroelectric power emerged as the most widely used renewable energy source, accounting for over 18% of the world's total power generation capacity in 2019.

The first hydroelectric power plant, a testament to the pioneering work of Nikola Tesla and George Westinghouse, was established in 1896 at Niagara Falls, marking a significant milestone in global electrification. The Niagara Falls power plants, capable of producing up to 4.9 million kW – enough to power 3.8 million homes – were eventually decommissioned due to their generation of electricity at 25 Hz, incompatible with modern standards (Fig. 2.13).

The potential energy in hydroelectric systems is determined by the flow or fall of water. As water is channeled through a pipe intake, its pressure drives turbine blades, spinning an electrical generator and producing electricity. This electricity is then transmitted to the grid. In storage systems, water is held in reservoirs created by dams and released to meet high electricity demands (Fig. 2.13). However,

Fig. 2.13 Adam's Power Station with Three Tesla AC Generators at Niagara Falls, 1896. *Source:* Coleman 1896/Public Domain

hydroelectric power can impact the environment, such as disrupting natural habitats and fish migration.

2.2.3.1 Advantages of Hydroelectricity

- Hydroelectricity is a sustainable energy source, as it does not consume water, returning it to its source.
- Capable of continuous operation, hydroelectric power can be generated 24/7, assuming the water source remains constant.
- It is a clean energy source, producing no waste by-products during electricity generation.
- Hydroelectric plants can regulate water usage, conserving it for periods of higher demand (Fig. 2.14).

2.2.3.2 Disadvantages of Hydroelectricity

- Potential environmental impacts, including disruption of animal habitats and extensive land use.
- High initial construction costs for dams and power plants.
- Expensive infrastructure for transmitting electricity to consumers.
- Dependence on consistent water supply and access.

Fig. 2.14 A hydroelectric power plant known as the Kruonis Pumped Storage Plant near Kruonis, Lithuania. *Source:* With Permission of Siemens Energy

2.2.4 Biomass Energy

Biomass energy, derived from organic materials like plants and animals, is a versatile and sustainable energy source. The sun's energy, captured by plants through photosynthesis, is stored as chemical energy. This energy can be released through combustion or converted into other forms like methane gas, ethanol, biodiesel, and biogases. The conversion processes, including anaerobic digestion, gasification, and fermentation, occur in biorefineries, akin to oil refineries and petrochemical plants. The biomass industry globally utilizes a range of conversion processes tailored to the available feedstock in each region, such as grasses, forestry by-products, agricultural residues, manure, and municipal solid wastes. Wood remains the most prevalent form of biomass used (Fig. 2.15).

Biomass energy applications fall into three categories: biopower, biofuels, and bioproducts. Biopower involves generating heat and electricity by directly burning biomass or converting it into biogases or liquid fuels. Wood, including compressed wood pellets made from wastes like sawdust and agricultural residues, is a common fuel source. These pellets, with their high energy density and ease of storage and transport, offer an efficient way to utilize by-products from forestry and agriculture. Landfills, rich in organic matter, also present opportunities for biopower as decomposing waste releases methane, which can be harnessed for electricity generation. The residual sludge from gasification is a valuable by-product, often sold as fertilizer.

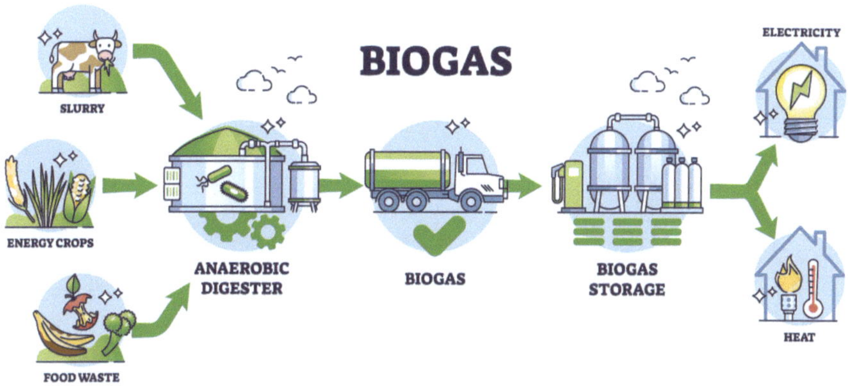

Fig. 2.15 Biogas Production Diagram. *Source:* Adobe Stock Photos

Biofuels, such as ethanol and biodiesel, are liquid fuels used for transportation. Ethanol, produced through fermentation of glucose extracted from crops like corn, wheat, and beets, can be blended with gasoline. Woody crops and switchgrass are emerging as new sources for ethanol production. Biodiesel, distilled from vegetable oils and animal fats, including recycled restaurant grease, is a biodegradable fuel that reduces emissions and can be used in diesel engines (Fig. 2.16).

The use of biomass energy has significant ecological and economic benefits. It is carbon-neutral, as the CO_2 released during combustion or conversion equals the amount absorbed during photosynthesis. Biomass has minimal sulfur content,

Fig. 2.16 Field for Biofuel Production. *Source:* Adobe Stock Photos

Fig. 2.17 Landfill – Municipal Waste Biofuel Source. *Source:* Aonprom Photo/Adobe Stock Photos

reducing acid rain potential. It supports agriculture and forestry industries by utilizing crop and wood residues. Biomass energy also optimizes land use, as feedstock crops can grow on nonarable land. Additionally, bioproducts can replace petroleum-based products, contributing to efficient waste management, especially in urban areas (Fig. 2.17).

Biomass energy harnesses the sun's energy stored in organic matter, offering a sustainable and versatile energy solution. It plays a crucial role in reducing greenhouse gas emissions and supports a circular economy by recycling waste into valuable energy and products.

2.2.5 Biomass Energy from Landfills and Biofuels

Biofuels, a renewable energy source, are derived from organic biomass such as trees, corn, grass, and agricultural or food wastes. These fuels are produced from plants that regrow annually, offering a sustainable energy alternative. Biofuels encompass a diverse range of fuels derived from organic matter. The rising costs of gas and oil have heightened interest in generating energy from plant fermentation.

Ethanol, the most well-known biofuel, is produced by fermenting sugars found in grains like corn, sorghum, barley, sugar cane, and sugar beets. Biodiesel, another biofuel variant, is derived from vegetable oils, animal fats, or recycled lubricants. Biogas, or landfill gas, is produced during anaerobic digestion, where bacteria break down organic materials in oxygen-free environments.

2.2.5.1 Benefits of Biofuels

- Biofuels can reduce carbon and sulfur emissions by 50–60%, contributing to decreased pollution and conservation of exhaustible resources.
- They are carbon-neutral: the carbon released during combustion is offset by the carbon absorbed during plant growth.
- Biofuels can be utilized without significant modifications to existing internal combustion engine systems.
- They reduce reliance on foreign oil suppliers.

2.2.5.2 Disadvantages of Biofuels

- Biofuels are not entirely clean energy sources. Their production and combustion still contribute to carbon emissions, partly due to the machinery used in cultivation and processing.
- The use of crops for biofuel production can compete with food production, raising ethical concerns about prioritizing energy over food.

2.2.5.3 Summary of Hay as a Biofuel

- Composed of lignin, cellulose, and hemicellulose.
- After compression, it has a density comparable to oak.
- Offers 5% higher energy content than oak.
- Simple to store and stack.
- Does not contribute additional greenhouse gases to the atmosphere.

An often-overlooked biofuel source is hay. Research by J. Gatterman and F. Overmoyer highlights the potential of using waste chaff, moldy, and fresh hay as a renewable energy source. Compressing hay into bricks could offer an efficient alternative to traditional firewood for home heating and industrial energy production. Each 40-pound bale of hay has the energy equivalent of approximately 2.2 gal of heating oil. This biofuel could be utilized in wood stoves, industrial boilers, and heating plants, though it may produce a significant amount of ash.

Biofuels present a viable alternative to fossil fuels, offering environmental benefits while posing certain challenges. The exploration of diverse sources like hay for biofuel production expands the potential of this renewable energy sector.

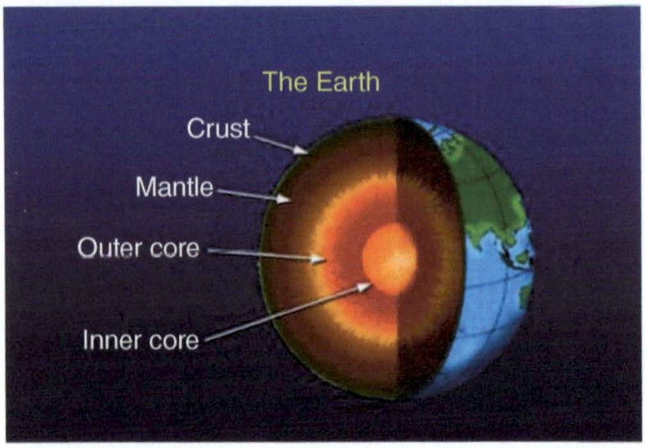

Fig. 2.18 Earth's Crust as a Source of Geothermal Energy

2.2.6 Geothermal Energy: Harnessing the Earth's Heat

Geothermal energy, derived from the Earth's internal heat, is a significant renewable resource. It originates from the planet's underground activity, particularly in regions with recent volcanic or seismic activity. These areas, where the Earth's crust is thin or disrupted, offer accessible sources of geothermal energy (Fig. 2.18).

2.2.6.1 Types of Geothermal Resources

Hydrothermal Systems: These systems occur where groundwater seeps into the Earth along fault lines and becomes heated by the underlying hot rock. Deep below the surface, high-pressure conditions can heat water to temperatures well above the boiling point at the surface. This heated water, moving through rock fractures, forms reservoirs and may surface naturally. Temperatures in these reservoirs range from 250 to 600 °F. In binary plants, hot water is used to boil a working fluid, typically an organic compound with a low boiling point, which then powers turbines to generate electricity. The water is subsequently reinjected into the ground for reheating.

Geo-pressured Zones: These are areas where saltwater or brine is trapped between layers of hot, impermeable rock. The brine, reaching temperatures of 200–400 °F, becomes highly pressurized and can be utilized for electricity production. Sometimes, large quantities of dissolved methane in the brine enhance the energy potential.

Hot and Dry Rocks: Found globally, these rocks represent a vast geothermal resource. Man-made wells are drilled, and water is pumped under high pressure to create fractures in the heated rocks, forming hydrothermal reservoirs. The heated water is then used for electricity generation.

Magma: Originating from the Earth's mantle and lower crust, magma is the source of volcanic activity. In some regions, magma reservoirs are closer to the surface and can be tapped for geothermal energy. Extracting energy from magma requires advanced drilling technology capable of withstanding extreme temperatures. Water is used to solidify the magma, acting as a heat exchanger for electricity generation. Another approach involves mixing water with iron oxide in the magma to produce hydrogen.

2.2.6.2 Natural Hydrothermal Features

Hydrothermal systems are also the source of natural hot springs and geysers, like the Castle Geyser in Yellowstone National Park (Fig. 2.19). These geothermal features are often associated with volcanic activity. Geysers, which intermittently eject steam and water, can be harnessed for both recreational and energy-generating purposes.

2.2.6.3 Advantages of Geothermal Energy

- Renewable and sustainable energy source.
- Can provide continuous power.
- Low environmental impact compared to fossil fuels.

Fig. 2.19 Castle Geyser Eruption, Yellowstone National Park, 2019. *Source:* Shaul Hurwitz/USGS/Public Domain

2.2.6.4 Challenges of Geothermal Energy

- Requires specific geological conditions.
- High initial investment for infrastructure development.
- Potential environmental impact from drilling and water usage.

Geothermal energy offers a promising alternative to traditional energy sources, leveraging the Earth's internal heat for sustainable power generation. Its development and utilization, however, require careful consideration of geological conditions and environmental impacts.

2.2.7 Geothermal Heating: Tapping into Earth's Stable Temperatures

Geothermal heating systems harness the Earth's consistent underground temperature, approximately 55 °F year-round, to heat and cool buildings. This method involves a ground loop system, where a network of pipes is buried underground. These pipes circulate a mixture of water and environmentally safe antifreeze, functioning as a large heat exchanger for the geothermal system.

2.2.7.1 How Geothermal Heating Works

- In colder months, traditional air source heat pumps work strenuously to extract heat from the cold air to warm buildings.
- Geothermal heat pumps, however, operate more efficiently by absorbing heat from the ground through the water circulating in the underground loop.
- This system is significantly more energy-efficient due to the constant underground temperature and the abundant heat stored beneath the surface.

2.2.7.2 Components of Geothermal Heat Pump Systems

- *Ground Heat Exchanger*: A loop of pipes buried near the building, through which a water-antifreeze mixture circulates, absorbing or releasing heat from/to the ground.
- *Heat Pump Unit*: In winter, the pump extracts heat from the ground to warm the building. In summer, it reverses the process, removing heat from the building and dissipating it into the ground.
- *Ductwork*: Delivers the heated or cooled air throughout the building.

2.2.7.3 Efficiency Comparison

- Traditional gas furnaces, which burn natural gas for heating, have an efficiency of up to 98%.
- Geothermal systems, by contrast, achieve efficiencies of 400–600% by utilizing the Earth's heat, making them a more energy-efficient option (Fig. 2.20).

Fig. 2.20 Nesjavellir Geothermal Power Plant in Pingvellir, Iceland. *Source:* Gretar Ívarsson/Wikimedia Commons/Public Domain

2.2.7.4 Advantages of Geothermal Heating

- Utilizes stable underground temperatures for efficient heating and cooling.
- More energy-efficient compared to traditional heating methods.
- Environmentally friendly, reducing reliance on fossil fuels.

2.2.7.5 Challenges

- Initial installation costs can be high due to the need for underground piping.
- Suitability depends on specific geological conditions.

While geothermal heating's initial setup costs can be significant, the long-term energy savings and environmental benefits make it a compelling option for sustainable building management.

2.2.8 Ocean Energy: Harnessing the Power of Tides

Ocean energy, particularly tidal power, is generated by the surge of ocean waters during the rise and fall of tides. This form of energy, influenced by the gravitational forces of the moon and sun along with Earth's rotation, has been utilized for over a millennium, notably in Europe for operating grain mills. Tidal energy is a consistent and pollution-free source, offering predictability and sustainability.

2.2.8.1 Tidal Energy Mechanisms

- *Tidal Barrages*: These structures stretch across inlets, featuring gates that regulate the flow of ocean water into a tidal basin during high tides and release it during low tides. The flowing water powers turbines within the basin to generate electricity.
- *Tidal Fences*: Installed in channels, these consist of vertical-axis turbines mounted within a fence-like structure, channeling all tidal water through the turbines.
- *Wave Energy Systems*: These systems capture the energy of ocean waves, intensified as they approach shorelines, and channel them through turbines.
- *Ocean Thermal Energy Conversion (OTEC)*: This technology leverages the temperature differential between warmer surface water and cooler deep water to produce electricity, though it faces economic challenges due to the cost of transporting electricity to land.

2.2.8.2 Environmental Impact and Global Presence

- Tidal power stations can affect local ecosystems, particularly in estuarine areas, by altering tidal levels and increasing turbulence. They may also impact navigation and recreational activities.
- Notable tidal power plants include South Korea's Sihwa Lake Station (254 MW), France's La Rance facility (240 MW) (Fig. 2.21), and Canada's Annapolis Royal Station (20 MW). The United States has potential sites for tidal power development, particularly in Alaska and Maine.

Fig. 2.21 Barrage of the Tidal Power Plant at La Rance River in Bretagne, France. *Source:* Dani 7C3/Wikimedia Commons/CC BY 2.5

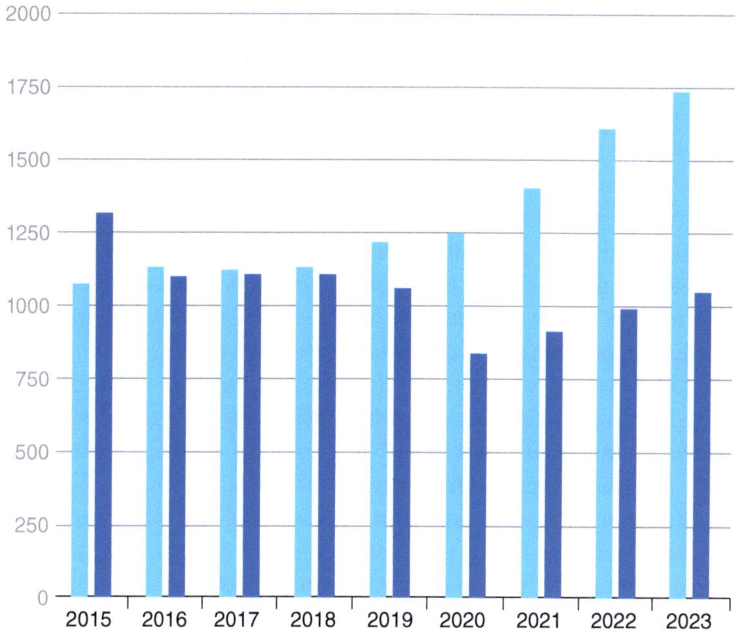

Fig. 2.22 Global Energy Investment in Clean Energy and in Fossil Fuels, 2015–2023. *Source:* International Energy Agency/https://www.iea.org/data-and-statistics/charts/global-energy-investment-in-clean-energy-and-in-fossil-fuels-2015-2023, last accessed 26 March 2024/CC BY 4.0

2.2.8.3 Global Renewable Energy Trends

- Renewable energy development is a focus in at least 48 countries, driven by economic and environmental considerations.
- Investments in renewable energy reached over US$30 billion in 2004, creating direct employment for millions.
- The share of renewables in global power usage has been steadily increasing, reaching 28.1% between 2020 and 2021 (Fig. 2.22).

2.3 Nuclear Energy: An Alternative Power Source

Nuclear power, functioning as an alternative energy source, harnesses controlled and continuous nuclear fission. While often categorized as nonrenewable due to the finite nature of its fuel sources, nuclear energy itself is renewable. However, the materials used in nuclear reactors, such as Uranium-235, are not renewable, complicating its classification. Thus, nuclear energy is best viewed as a hybrid energy source.

In nuclear power plants, the heat generated from nuclear reactions is converted into steam, which then powers turbines to produce electricity. Nuclear power contributes about 6% of global energy and 16% of the world's electricity. In the United States, nuclear power plants generated 772 billion kWh of electricity in 2022, accounting for 47% of the nation's carbon-free electricity, making it the largest domestic source of clean energy.

To date, the primary form of nuclear energy in use is derived from the fission of heavy nuclei like Uranium-235 or Plutonium-239. Uranium, a naturally radioactive element, is found globally, while Plutonium-239 is primarily used in breeder reactors. These reactors are advantageous as they produce fissionable material but have faced political opposition due to complex safety concerns.

Nuclear reactors, however, come with significant drawbacks. The major concern is the radioactivity of fissionable isotopes, which pose severe health risks in the event of accidental atmospheric exposure. The Chernobyl disaster in 1989 is a stark reminder of these risks. Additionally, spent nuclear fuel rods remain radioactive for hundreds of thousands of years, necessitating secure disposal solutions like the Yucca Mountain facility in Nevada. Objects in close proximity to the reactor core also become mildly radioactive and require careful handling.

In research laboratories, significant attention is given to the fusion of hydrogen isotopes. "Cold fusion," a process that would offer an environmentally friendly and virtually limitless energy source, remains a scientific aspiration. Despite numerous reports of successful cold fusion experiments, replicable results have yet to be published.

2.3.1 Advantages and Disadvantages of Nuclear Power

2.3.1.1 Advantages of Nuclear Power

- Enhanced safety due to new technology.
- Zero greenhouse gas emissions from nuclear power plants.
- Safe, long-term underground storage of nuclear waste.
- Low operational costs; a doubling of uranium costs would only increase plant operating costs by about 7%.

2.3.1.2 Disadvantages of Nuclear Power

- Public apprehension regarding potential plant disasters and limited nuclear waste storage options.
- High initial and maintenance costs.
- Generation of highly toxic radioactive waste, posing health risks.
- Negative publicity from incidents like Chernobyl and the Fukushima disaster.

Enrichment of Uranium-235 from the more abundant Uranium-238 is necessary for nuclear fission. Figure 2.23 illustrates the distribution of nuclear power production across various countries. Global uranium reserves are estimated to last approximately 50 years.

Fig. 2.23 India's Kakrapar Atomic Power Nuclear Plant. *Source:* DAE (GODL-India)/ Wikimedia Commons/Public Domain

Electricity production, primarily from fossil fuels, is the leading source of CO_2 emissions worldwide. With global energy demands projected to double by 2050, nuclear power offers an efficient, low-carbon, and controllable energy production alternative. Despite the challenges, nuclear energy remains a crucial component in reducing reliance on fossil fuels and addressing future energy needs.

2.3.2 Nuclear Fission and Uranium Enrichment

To access fissionable Uranium-235, it must be enriched from the more prevalent Uranium-238, typically found in minerals like pitchblende. Figure 2.24 illustrates the proportion of nuclear power production across various nations. Current global uranium reserves are projected to last approximately 50 years.

Electricity generation, primarily reliant on fossil fuels, is the leading contributor to global CO_2 emissions. With the world's energy demands expected to double by 2050, renewable energy, despite its numerous benefits, cannot singularly meet the energy requirements of an estimated 8 billion people. Nuclear power, leveraging controlled and continuous nuclear fission, emerges as a viable alternative. It offers an efficient, low-carbon, and manageable energy production method. The heat generated from nuclear reactions is converted into steam, driving turbines to produce electricity.

Currently, nuclear fission, the process of splitting heavy atoms to sustain the power grid, is the prevalent method in nuclear power stations. This technology, while effective in energy production, generates radioactive waste that remains

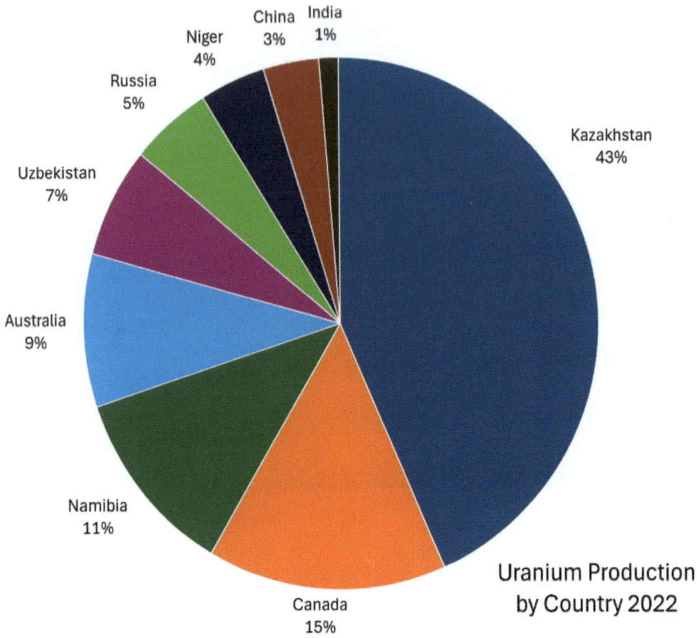

Fig. 2.24 Percentage of Uranium produced by countries in 2021. *Source:* Data was sourced from the World Nuclear Association https://world-nuclear.org/information-library/facts-and-figures/uranium-production-figures

hazardous over extended periods, posing significant health risks and challenges in waste management and storage.

2.3.3 Fusion: The Future of Nuclear Energy

Nuclear fusion, a process that mirrors the Sun's energy production, represents a significant advancement in nuclear energy. A notable breakthrough occurred on December 13, 2022, when the Lawrence Livermore National Laboratory (LLNL) in California reported producing more energy from a fusion experiment than was input. This achievement, while still in its nascent stage, opens the door to the potential of fusion as a sustainable energy source.

Fusion differs fundamentally from nuclear fission, the traditional method of nuclear energy production. Fusion offers a higher energy yield with minimal and short-lived radioactive waste. Importantly, it does not contribute to greenhouse gas emissions, making it a promising option for addressing climate change concerns.

The concept of fusion involves heating and compressing hydrogen atoms under extreme conditions to initiate the fusion process. This method promises

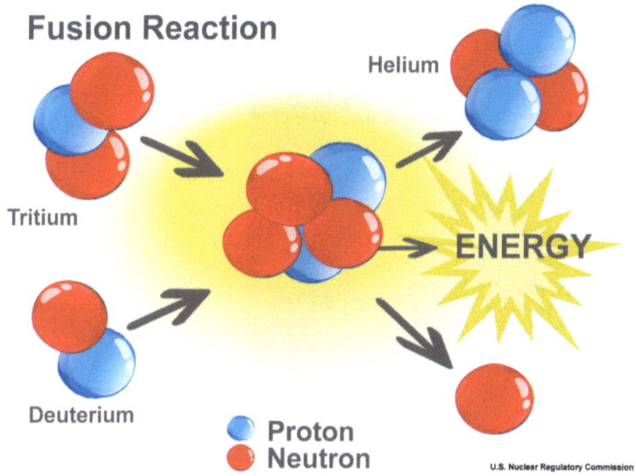

Fig. 2.25 Fusion Model. *Source:* Nuclear Regulatory Commission. https://www.flickr.com/photos/nrcgov/52827180731/

a clean, virtually limitless energy source, aligning with global energy sustainability goals. However, the transition from experimental success to practical, scalable energy production presents significant scientific and engineering challenges, including the need for advanced technology and substantial financial investment.

The potential of fusion energy is vast, with the promise of providing a significant portion of the world's energy needs without the environmental drawbacks associated with fossil fuels or traditional nuclear fission. The pursuit of fusion energy is driven by the prospect of a reliable, carbon-neutral energy source that could revolutionize how we generate power (Fig. 2.25).

While the timeline for fusion energy becoming a mainstream power source remains uncertain, the ongoing research and development in this field are critical steps toward a sustainable energy future. The continued exploration and refinement of fusion technology hold the key to unlocking this powerful and clean energy source (Fig. 2.26).

The journey toward practical fusion energy involves overcoming technical hurdles and enhancing the efficiency and output of the fusion process. The success of experiments like those at LLNL serves as a catalyst for further innovation in this exciting field of energy research (Fig. 2.27).

The preamplifier module amplifies the laser energy *en route* to the Target Chamber (Fig. 2.28).

The Target Chamber at LLNL's National Ignition Facility witnessed the delivery of over 2 million joules of ultraviolet energy by 192 laser beams to a tiny fuel pellet, achieving fusion ignition on December 5, 2022.

Fig. 2.26 LLNL Hohlraum. *Source:* John Jett and Jake Long/Lawrence Livermore National Laboratory/Public Domain

Fig. 2.27 LLNL Ignition Facility preamplifier. *Source:* LLNL

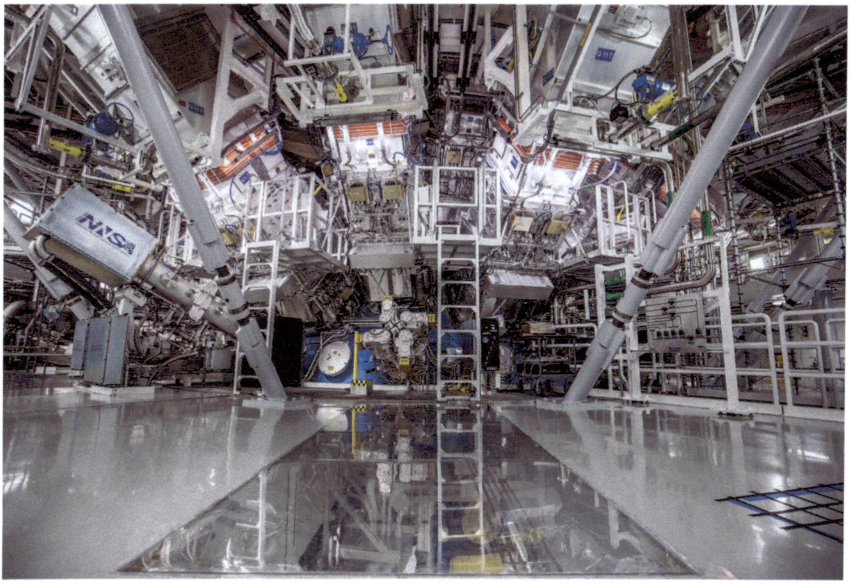

Fig. 2.28 LLNL Target Chamber. *Source:* Jason Laurea/Lawrence Livermore National Laboratory/Public Domain

3

Hydrogen Properties

3.1 General Characteristics and Physical Properties of Hydrogen

Hydrogen, derived from the Greek words "hydro" (water) and "genes" (forming), and known in Latin as hydrogenium, is the first element in the periodic table, identified by the atomic number 1. English chemist and physicist Henry Cavendish officially recognized it as an element in 1766, although its unique gaseous properties were earlier observed by Robert Boyle in 1671 through his experiments involving iron and diluted hydrochloric acid (list of scientists, contributed to hydrogen research presented in Fig. 3.1).

This simplest of elements is composed of a positively charged nucleus orbited by a single electron, represented by the electron configuration $_1s1$. Hydrogen is the most prevalent element in the universe, constituting 90% of its mass. Despite its abundance, it rarely exists in its elemental form on Earth, as it readily bonds with other elements. It is also the lightest element, with a density of 0.08988 g/L at standard atmospheric pressure.

Hydrogen is a colorless, odorless, tasteless, and nonpoisonous gas under normal conditions on Earth. It typically exists as a diatomic molecule, meaning each molecule has two atoms of hydrogen; this is why pure hydrogen is commonly expressed as "H_2" (see Fig. 3.2).

- The atomic structure of hydrogen, with a single electron orbiting a positively charged nucleus, renders each atom highly reactive. This leads to its natural occurrence only in the form of hydrogen molecules (H_2).
- The energy within a single hydrogen molecule is less than the total energy of two separate hydrogen atoms, amounting to 436 kJ/mol.

Hydrogen Energetics: A Future Clean Energy Carrier, First Edition. Roman J. Press.

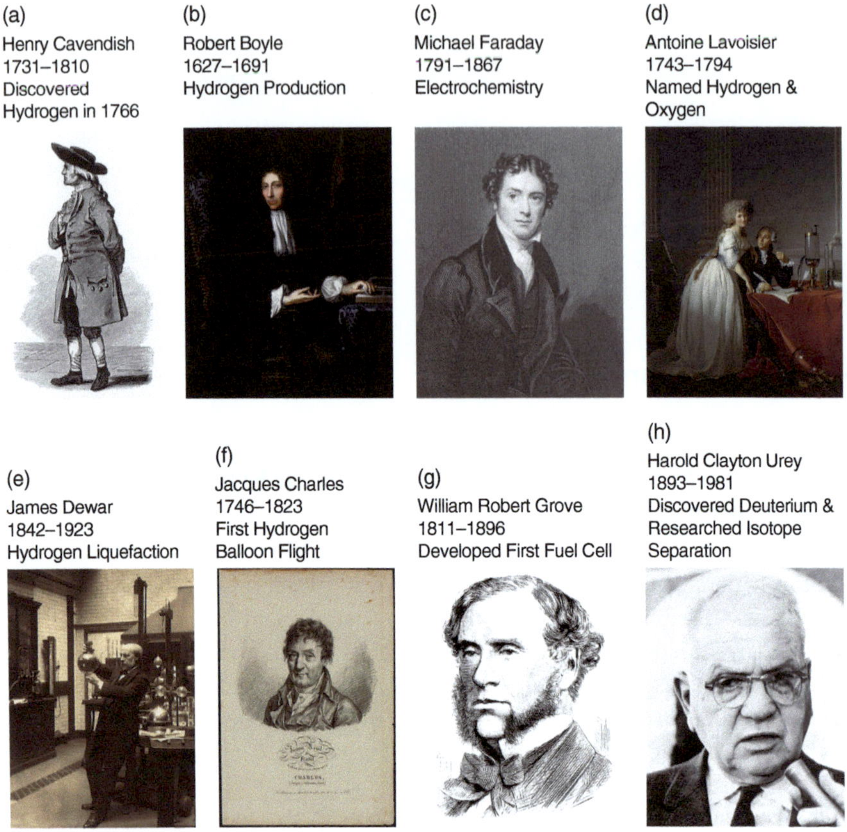

Fig. 3.1 Scientists Contributed to Hydrogen Research. *Source:* (a) Archivist/Adobe Stock Photos; (b) Johann Kerseboom/Wikimedia Commons/Public Domain; (c) Georgios Kollidas/Adobe Stock Photos; (d) Jacques-Louis David/Wikimedia Commons/Public Domain; (e) Thomas Thorpe/Wikimedia Commons/Public Domain; (f) Tissandier Collection/The Library of Congress/Public Domain; (g) Unknown Author/Wikimedia Commons/Public Domain; (h) NASA/Public Domain

Fig. 3.2 Hydrogen Molecule Model. *Source:* Adobe Stock

- Hydrogen is also the lightest element, about 1/15th as heavy as air, causing it to rise and dissipate rapidly. The primary physical properties of hydrogen are detailed in Table 3.1.
- It exhibits a high capacity for adsorption, readily attaching to certain substances, and is highly soluble in water, alcohol, and ether.
- Under extremely low pressure, close to a vacuum, only individual hydrogen atoms exist, as there is no opportunity for them to combine. When cooled to its boiling point of −252.7 °C, hydrogen becomes a transparent, odorless liquid. Liquid hydrogen is significantly lighter than water, noncorrosive, highly reactive, and expands approximately 840 times when converted from liquid to gas.
- The majority of a hydrogen atom's mass is concentrated in its nucleus. The proton is more than 1,800 times more massive than the electron, and the mass of neutrons is almost equivalent to that of protons.

The radius of the electron orbit is about 100,000 times larger than that of the nucleus. Consequently, a hydrogen atom is mainly composed of empty space.

Table 3.1 Physical Properties of Hydrogen

Property	Value
Molecular Weight	
Molecular weight	2.016 g/mol
Solid Phase	
Melting point	−259 °C
Latent heat of fusion (1.013 bar, at triple point)	58.158 kJ/kg
Liquid Phase	
Liquid density (1.013 bar at boiling point)	70.973 kg/m^3
Liquid/gas equivalent (1.013 bar and 15 °C)	844 v/v
Boiling point (1.013 bar)	−252.8 °C
Latent heat of vaporization (1.013 bar at boiling point)	454.3 kJ/kg
Critical Point	
Critical temperature	−240 °C
Critical pressure	12.98 bar
Critical density	30.09 kg/m^3

(Continued)

Table 3.1 (Continued)

Property	Value
Triple Point	
Triple-point temperature	−259.3 °C
Triple-point pressure	0.072 bar
Gaseous Phase	
Gas density (1.013 bar at boiling point)	1.312 kg/m^3
Gas density (1.013 bar and 15 °C)	0.085 kg/m^3
Compressibility factor (*Z*) (1.013 bar and 15 °C)	1.001
Specific gravity (air = 1) (1.013 bar and 21 °C)	0.0696
Specific volume (1.013 bar and 21 °C [70 °F])	11.986 m^3/kg
Heat capacity at constant pressure (Cp) (1 bar and 25 °C)	0.029 kJ/(mol K)
Heat capacity at constant volume (Cv) (1 bar and 25 °C)	0.021 kJ/(mol K)
Ratio of specific heats (Gamma:Cp/Cv) (1 bar and 25 °C)	1.384259
Viscosity (1.013 bar and 15 °C)	0.0000865 Poise
Thermal conductivity (1.013 bar and 0 °C)	168.35 mW/(m K)
Others	
Solubility in water (1.013 bar and 0 °C)	0.0214 v/v
Concentration in air	0.00005 vol%
Autoignition temperature	560 °C
Diffusion	1.697 m^2/h
Flame temperature	2318 °C
Flammable range	4–74% by vol in air
Heat of combustion by mass	28 670 kcal/kg
Ignition energy	0.02 mJ
Volumetric energy density	57.8 kcal/kg mol
Heat of fusion (H_2)	0.117 kJ/mol

3.1.1 Chemical Properties of Hydrogen

Hydrogen possesses several key chemical properties that significantly influence its application as a fuel source:

- It reacts with oxygen to produce water, an essential compound for life on Earth.
- Hydrogen boasts a high energy content by weight, nearly triple that of gasoline. However, its energy density per volume is relatively low under standard

temperature and pressure. This density can be enhanced by compressing hydrogen or by liquefying it at extremely low temperatures. Additionally, hydrogen can be absorbed into metal hydrides for storage.

- Notably flammable, hydrogen ignites easily and burns across a broad flammability range, from 4% to 74% concentration in air.
- Hydrogen combustion produces a pale-blue, nearly invisible flame, which can make hydrogen fires hard to detect visually.
- When hydrogen burns, it does not emit carbon dioxide (CO_2), particulates, or sulfur. However, under certain conditions, it may produce nitrous oxide (NO_x) emissions.
- Hydrogen can be generated from renewable sources, including the reforming of ethanol (a process that emits some CO_2) and through the electrolysis of water.
- Hydrogen is capable of forming compounds with all elements except the noble gases.
- With an electronegativity of 2.2, hydrogen can behave as a nonmetallic element, forming ionic salts known as hydrides (e.g., NaH and CaH_2) where hydrogen exists as H-ions. Alternatively, it can act more metallic, forming covalent bonds with nonmetals like sulfur, nitrogen, and halogens.
- Hydrogen is a component of strong acids such as HCl, H_2SO_4, and HNO_3. In these compounds, the H+ ion, essentially a bare nucleus, strongly attracts electrons.
- Hydrogen also forms a wide array of organic compounds, including hydrocarbons like methane (CH_4), ethane (C_2H_6), propane (C_3H_8), and butane (C_4H_{10}) as well as alcohols like methanol (CH_3OH) and ethanol (C_2H_5OH). These organic compounds can also include acids, amines, and various compounds containing heteroatoms like sulfur, halogens, and phosphorus.

3.1.2 Chemical Reactions of Hydrogen

Hydrogen engages in various chemical reactions, showcasing its versatility:

1) *Reaction with Alkali and Alkaline Earth Metals*: Hydrogen gas directly reacts with highly reactive metals from groups IA and IIA of the periodic table, forming metal hydrides:

$$H_2(g) + 2Na(I) \rightarrow 2NaH(s)$$

$$H_2(g) + Ca(1) \rightarrow CaH_2(s)$$

2) *Reduction of Metal Oxides*: Hydrogen can reduce the oxides of several metals, such as silver, copper, lead, bismuth, and mercury, to their

elemental forms. This reaction is industrially significant for producing metals like tungsten:

$$3H_2(g) + 2WO_3(s) \Rightarrow 2W(s) + 3H_2O(g)$$

3) *Formation of Halogen Halides*: Hydrogen reacts with halogens to form halogen halides:

$$H_2(g) + Cl_2(g) \rightarrow 2HCl(g)$$

4) *Ammonia Production*: The catalytic reduction of nitrogen by hydrogen is a cornerstone of industrial ammonia production, crucial for fertilizers. This process, known as the Haber process, operates at high temperatures (450–500 °C) and pressures (35,000–40,000 kPa):

$$3H_2(g) + N_2(g) \Leftrightarrow 2NH_3(g)$$

5) *Hydrogenation of Unsaturated Hydrocarbons*: Atomic hydrogen reacts with unsaturated hydrocarbons, like alkenes and alkynes, to form saturated alkanes. For instance, hydrogen adds to ethylene to produce ethane:

$$H_2 + H_2C = CH_2 \rightarrow H_3C - CH_3$$

6) *Combustion to Form Water*: Hydrogen reacts with oxygen to form water, a reaction fundamental to its energy applications:

$$H_2(g) + O_2(g) \rightarrow 2H_2O(g)$$

7) This reaction is notably slow at room temperature but can be explosively accelerated by a catalyst like platinum or a single electric spark. Hydrogen's ability to combust in air is one of its most significant properties.

3.1.3 Health Effects of Hydrogen

Hydrogen, while not inherently toxic, poses several risks and hazards:

- *Extreme Flammability*: Hydrogen is highly flammable when mixed with air or oxygen. Exposure to heat can lead to violent combustion or explosions. Its combustion in air produces a pale blue, nearly invisible flame, making hydrogen fires hard to detect.

- *Reactivity with Other Elements*: Hydrogen reacts vigorously with halogens and strong oxidants, potentially causing fires and explosions. This reaction is further intensified in the presence of metal catalysts like platinum and nickel.
- *Risk of Suffocation*: Although hydrogen is not toxic, it can cause suffocation by reducing the oxygen concentration in the air to levels insufficient for sustaining life.
- *Oxygen Deprivation*: Inhalation of hydrogen can lead to oxygen deprivation, resulting in symptoms like headaches, dizziness, drowsiness, unconsciousness, nausea, vomiting, and depression. In high concentrations, hydrogen can be fatal.
- *Invisibility and Odorlessness*: Hydrogen's lack of color and odor means there is no immediate indication of dangerous concentrations in the air. It is crucial to monitor hydrogen levels using appropriate gas detectors to ensure safety.
- *High Combustion Energy*: Hydrogen has the highest energy release per unit weight among commonly occurring materials, which contributes to its potential hazards.

3.1.4 Hydrogen Isotopes

Hydrogen is the only element that has different names for its isotopes (see Fig. 3.3). There are three major hydrogen isotopes:

- ${}^{1}_{1}H$ protium 1 proton and 0 neutrons,
- ${}^{2}_{1}H$ deuterium 1 proton and 1 neutron,
- ${}^{3}_{1}H$ tritium (radioactive) 1 proton and 2 neutrons.

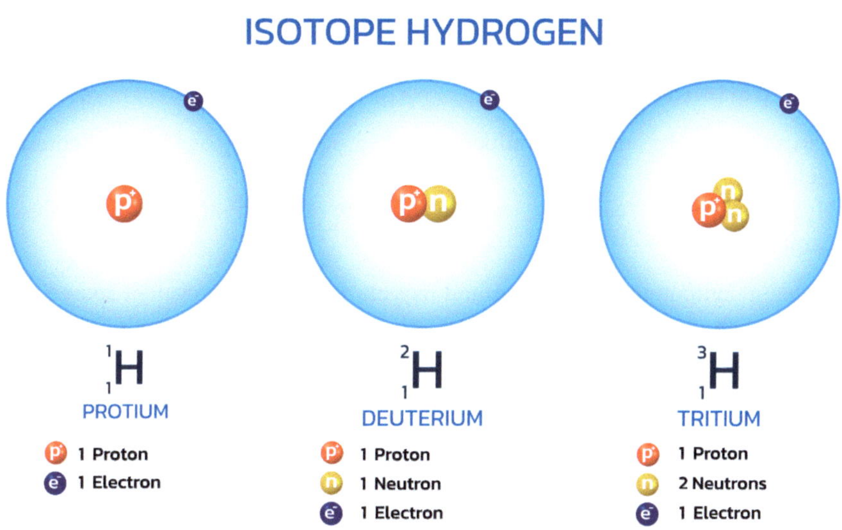

Fig. 3.3 Hydrogen Isotopes. *Source:* Adobe Stock Photos

3.2 Hydrogen Bonding

Hydrogen bonding is a type of electrostatic attraction that occurs between a hydrogen (H) atom, which is covalently bonded to a more electronegative "donor" atom or group (Dn), and another electronegative atom that possesses a lone pair of electrons, known as the hydrogen bond acceptor (Ac). Common donor and acceptor atoms include water (H_2O), oxygen (O), ammonia (NH_3), and hydrogen fluoride (HF).

While hydrogen bonding is generally weaker than ionic or covalent bonding, it is significantly stronger than van der Waals forces. This type of bonding can occur both intermolecularly (between atoms of different molecules) and intramolecularly (within the same molecule). For instance, in water molecules, hydrogen is covalently bonded to the more electronegative oxygen atom. Hydrogen bonding in water results from dipole–dipole interactions between the hydrogen atom of one water molecule and the oxygen atom of another H_2O molecule.

3.3 Occurrence of Hydrogen

Hydrogen stands as the most abundant element in the universe, constituting about 90% of the atoms or 75% of the mass of the universe. On Earth, hydrogen accounts for approximately 1% of the planet's total mass. Despite its abundance, hydrogen is rarely found in its pure form on Earth; it mainly appears in volcanic gases and some natural gases. The scarcity of molecular hydrogen on Earth is attributed to two factors:

- Earth's gravity has minimal influence on hydrogen, being one of the lightest elements.
- Hydrogen's high reactivity leads it to form a wide array of chemical compounds.

Water, a compound of hydrogen and oxygen, is the most prevalent source of hydrogen on Earth, covering about two-thirds of the planet's surface. Hydrocarbons, comprising hydrogen and carbon, such as petroleum and natural gas, are other significant sources.

Air also contains water vapor, varying from about 6% in humid regions to about 0.1% in deserts. Water is vital for life and is present in all living organisms, constituting nearly 70% of the human body. Hydrogen is a fundamental component of all organic matter, including plant, animal, and fossil substances.

It is found in energy-providing foods like fats, proteins, and carbohydrates and is a major element in biomass, including DNA, the molecules encoding genetic information.

In stars and galaxies, hydrogen predominantly exists in a plasma state, including in our Sun. In stars, hydrogen nuclei undergo nuclear reactions to form helium atoms, generating the light and heat we receive from the Sun and other stars. Under extreme conditions, such as the high pressure at the centers of giant planets, hydrogen transforms into a metallic state, acting as an electrical conductor.

Metallic hydrogen, theorized in 1935 by Eugene Wigner and Hillard Bell Huntington, is believed to exist in a partially liquid state under high pressure and temperatures, potentially in abundance within the intensely hot and compressed interiors of Jupiter and Saturn as well as some exoplanets.

In Earth's atmosphere, the presence of hydrogen is minimal. The composition of dry air by mole fraction includes 78.08% nitrogen, 20.95% oxygen, 0.93% argon, 0.04% carbon dioxide, and 0.00005% hydrogen, along with trace amounts of other gases. At sea level, air typically contains around 1% water vapor.

3.4 Comparing Hydrogen to Other Fuels

When assessing the energy content of commonly used fuels, it is notable that 1 gal of gasoline is equivalent to 380 SCF (Standard Cubic Feet) of hydrogen (H_2) or 113 SCF of natural gas or 8.3 pounds of coal. This comparison highlights significant challenges in storing and transporting hydrogen compared to other fuels. Hydrogen's volume for a given energy amount is substantially larger than that of gasoline, natural gas, or other fuels with equivalent energy content. Table 3.2 illustrates this comparison across major physical parameters of different fuels.

The specific energy value of compressed gases and gasoline is presented in Table 3.3.

As indicated in the tables, hydrogen's performance per unit volume is less efficient compared to other fuel sources. For example, hydrogen at 1,000 psi (pounds per square inch) pressure delivers only about 2.5% of the energy content found in gasoline. At a pipeline pressure of 5,000 psi, hydrogen's energy content increases to approximately 114,285 BTU/ft^3, which is about 5 times higher than at 1,000 psi and 10 times higher at 10,000 psi. However, utilizing such high pressures in pipelines, coupled with the need for a larger volume of hydrogen, necessitates the use of pipelines with larger diameters. Additionally, it is important to consider the implications of these requirements on the infrastructure and overall feasibility of hydrogen as a fuel source.

Table 3.2 Comparative Fuel Properties

		Main Constituent	Boiling Point	Latent Heat of Vaporization	Specific Heating Value	Ignition Temperature	Lower ignition Limit	Upper ignition Limit
Substance	Density	%/wt	°C	kJ//kg	MJ/kg	°C	% by gas volume in air	% by gas volume in air
Regular Gasoline	0.715–0.765 kg/L	86 C 14 H	25–215	380–500	42.7	~300	~0.6	~8.0
Diesel Fuel	0.81 0.815–0.855 kg/L	86 C 13 H	180–360	~250	42.5	~250	~0.6	~6.6°
Crude Oil	0.70–1.0 kg/L	80–83 C 10–14 H	25–360	222–352	46.1	~220	~0.6	~6.5
Ethanol (C_2H_5OH)	0.79 kg/L	52 C 13 H 35 O	78	904	26.8	420	3.5	15
Methanol (CH_3OH)	0.79 kg/L	38 12 H 50 O	65	1110	19.7	450	5.5	26
Natural Gas	~0.83 kg/m^3	76 C 24 H	−162		47.7	–	–	–

Hydrogen (H_2)	0.09 kg/m^3	100 H	−253	120	560	4	77
Methane (CH_4)	0.72 kg/m^3	75 C 25 H	−162	50	650	5	15
Propane (C_3H_8)	2 kg/m^3	82 C 18 H	−43	46.3	470	1.9	9.5

Source: United States Department of Energy, https://afdc.energy.gov/fuels/properties/Public Domain

Table 3.3 Specific Energy Value of Compressed Gases and Gasoline

	BTU/ft³	BTU/ft³	BTU/ft³	BTU/ft³
Fuel	**at STP**	**1,000 psi**	**5,000 psi**	**10,000 psi**
Hydrogen	320	22,857	~114,300	~230,000
Methane	985	70,357		~680,000
Propane (gas)	2,450	~175,000		
Propane (non-compressible liquid)	782,000	782,000	782,000	782,000
Gasoline (non-compressible liquid)	894,740	894,740	894,740	894,740

4

Fuel Cells: Essential Information

4.1 Overview

Fuel cells (FCs) represent a significant shift from traditional combustion-based energy generation, as they produce electricity through an electrochemical process. This process involves converting the chemical energy of hydrogen and oxygen into electric current. Unlike combustion engines, FCs generate propulsion power without moving parts, avoiding combustion-related emissions and achieving higher efficiency. In contrast to batteries, which require hours for recharging, hydrogen used in FCs can be replenished in mere minutes.

The concept of FCs dates back to 1839, surprisingly predating the development of the internal combustion engine by Nicolaus August Otto in 1876. English lawyer and physicist Sir William Robert Grove conceived what he called the "Grove element" in that year. Interestingly, the foundational principle of the FC was established even before Otto unveiled his first internal combustion engine in Deutz, a suburb of Cologne. Shortly after Grove's discovery, German physicist Moritz Jacobi successfully used several Grove elements to power a boat electrically on the Neva River in St. Petersburg. However, the requirement for platinum in this method proved to be financially prohibitive, leading Jacobi to eventually abandon his project.

Grove's experiments effectively reversed the process of electrolysis, which had been used by his compatriot Anthony Carlisle 40 years earlier to split water into hydrogen and oxygen using an electric current.

It took over 120 years for the "fuel cell," as scientists later named the Grove element, to find a new application. In 1965, NASA's Gemini space capsules, part of its manned space flight program, utilized FCs to power radio communications and onboard computers. The advantages of these FCs included their functionality in weightless conditions, low weight, and the ability to use the generated water for astronauts in emergencies.

Hydrogen Energetics: A Future Clean Energy Carrier, First Edition. Roman J. Press.

The FCs used in the Gemini spacecraft were fundamentally similar to Grove's original design. Each FC comprising three basic components: the fuel electrode (anode), the oxygen electrode (cathode), and the electrolyte situated between them.

4.2 FC Technologies: Classification and Comparison

FCs are classified based on their anode/cathode reactions and the types of electrolytes they use, as detailed in Table 4.1. Modern FC development primarily focuses on hydrogen, methanol, or methane as fuel sources. Among these, hydrogen-powered FCs have seen significant advancements, with some models reaching commercial viability. The most advanced and practical among these is the polymer electrolyte membrane fuel cell (PEMFC), which employs an electrolyte to separate the anode and cathode electrodes.

4.2.1 Polymer Electrolyte Membrane Fuel Cells

Polymer Electrolyte Membrane Fuel Cells (PEMFCs), also known as proton exchange membrane fuel cells, stand out for their high power density, low weight, and compact size. These FCs use a solid polymer as an electrolyte and porous carbon electrodes containing a platinum or platinum alloy catalyst. They operate on hydrogen, oxygen from the air, and water, typically fueled by pure hydrogen from storage tanks. When hydrogen is derived from hydrocarbon fuels, a reactor is needed to reduce carbon monoxide in the fuel gas due to the platinum catalyst's sensitivity to carbon monoxide poisoning, adding complexity to the fuel preparation system.

PEMFCs operate at relatively low temperatures, around 60–100 °C (typically 80 °C or 176 °F), allowing for quick start-ups and enhanced durability of system components. At these temperatures, the water produced at the cathode is in both liquid and vapor forms. The electrochemical reactions in PEMFCs consist of two half-reactions: oxidation at the anode and reduction at the cathode. The overall cell reaction produces electric power, water, and surplus heat, with the water being expelled from the FC by airflow.

$$\text{Anode: } H_2 \rightarrow 2H^+ + 2e^-$$

$$\text{Cathode: } \tfrac{1}{2}O_2 + 2H^+ + 2e^- \rightarrow H_2O$$

$$\text{Cell: } H_2 + \tfrac{1}{2}O_2 \rightarrow H_2O$$

Table 4.1 Comparison of Fuel Cell Technologies

Fuel Cell Type	Common Electrolyte	Operating Temperature	Typical Stack Size	Electrical Efficiency (LHV)	Applications	Advantages	Challenges
Polymer Electrolyte Membrane (PEMFC)	Perfluorosulfonic acid	<120 °C	<1 to 100 kW	60% direct H_2; 40% reformed fuel	• Backup power • Portable power • Distributed generation • Transportation • Specialty vehicles	• Solid electrolyte reduces corrosion and electrolyte management problems • Low temperature • Quick start-up and load following	• Expensive catalysts • Sensitive to fuel impurities
Alkaline (AFC)	Aqueous potassium hydroxide soaked in a porous matrix or alkaline polymer membrane	<100 °C	1–100 kW	60%	• Military • Space • Backup power • Transportation	• Wider range of stable materials allows lower cost components • Low temperature • Quick start-up	• Sensitive to CO_2 in fuel and air • Electrolyte management (aqueous) • Electrolyte conductivity (polymer)

(Continued)

Table 4.1 (Continued)

Fuel Cell Type	Common Electrolyte	Operating Temperature	Typical Stack Size	Electrical Efficiency (LHV)	Applications	Advantages	Challenges
Phosphoric Acid (PAFC)	Phosphoric acid soaked in a porous matrix or imbibed in a polymer membrane	150–200 °C	5–400 kW, 100 kW module (liquid PAFC); <10 kW (polymer membrane)	40%	• Distributed generation	• Suitable for CHP • Increased tolerance to fuel impurities	• Expensive catalysts • Long start-up time • Sulfur sensitivity
Molten Carbonate (MCFC)	Molten lithium, sodium, and/or potassium carbonates, soaked in a porous matrix	600–700 °C	300 kW to 3 MW, 300 kW module	50%	• Electric utility • Distributed generation	• High efficiency • Fuel flexibility • Suitable for CHP • Hybrid/gas turbine cycle	• High-temperature corrosion and breakdown of cell components • Long start-up time • Low power density
Solid Oxide (SOFC)	Yttria-stabilized zirconia	500–1,000 °C	1 kW to 2 MW	60%	• Auxiliary power • Electric utility • Distributed generation	• High efficiency • Fuel flexibility • Solid electrolyte • Suitable for CHP • Hybrid/gas turbine cycle	• High-temperature corrosion and breakdown of cell components • Long start-up time • Limited number of shutdowns

Source: US Department of Energy/https://www.energy.gov/eere/fuelcells/articles/comparison-fuel-cell-technologies-fact-sheet, last accessed 26 March 2024/Public Domain

The half-reactions would typically proceed slowly; hence, catalysts are used at both the anode and the cathode to accelerate them. The use of a noble-metal catalyst (platinum) for the separation of hydrogen's electrons and protons contributes to the relatively high cost of the system. PEMFCs are particularly well suited for vehicle applications, such as cars, buses, and heavy-duty trucks, as well as for limited stationary applications.

4.2.1.1 PEMFC Fabrication Hardware

A PMFC's core component is the membrane electrode assembly (MEA), which comprises the membrane, anode and cathode electrodes, catalyst layers, and gas diffusion layers. The gaseous fuel is supplied to the anode, while air is fed to the cathode. Integral to the MEA are gaskets that seal the assembly to prevent gas leakage, and bipolar plates, which combine individual PEMFCs into a stack and facilitate the distribution of gaseous fuel and air (Fig. 4.1).

4.2.1.2 Membrane Electrode Assembly

The MEA is a critical integration of the anode, membrane, and cathode. Initially developed in the 1960s for the Gemini space program, these assemblies used 4 mg/cm^2 membrane area. Modern advancements have reduced the total platinum loading to

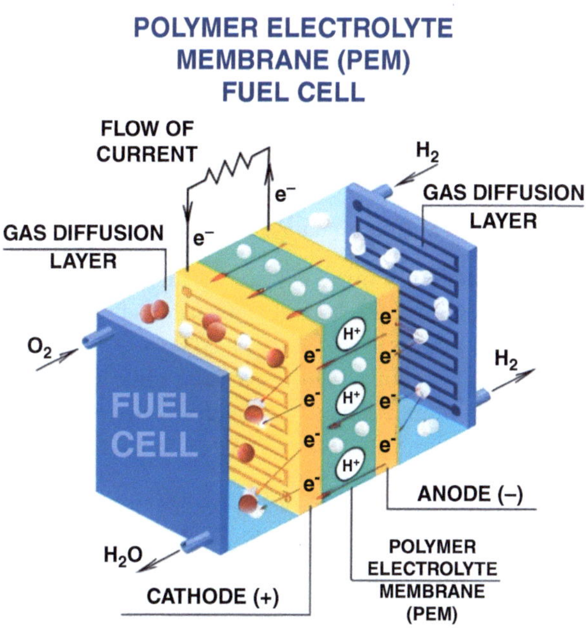

Fig. 4.1 Polymer Electrolyte Membrane Fuel Cell (PEMFC) Diagram. *Source:* Adobe Stock

approximately 0.5 mg/cm^2, with laboratory research achieving platinum loadings as low as 0.15 mg/cm^2. Each MEA in a FC stack is sandwiched between two bipolar plates, sealed with rubbery polymer gaskets to ensure a gas-tight enclosure.

This evolution in FC technology has significantly improved performance. For instance, the current generated per milligram of platinum has increased from about 0.5 A in the Gemini era to 15 A in contemporary designs. The catalyst layer's thickness depends on the amount of platinum used in each electrode. With about 0.15 mg Pt/cm^2, the catalyst layer is roughly 10 μm thick, less than half the thickness of a standard sheet of paper. When incorporated into well-engineered components like backing layers, flow fields, and current collectors, this MEA, with a total thickness of about 200 μm, can generate over half an ampere of current per square centimeter at a voltage of only 0.7 V between the cathode and the anode.

4.2.1.3 The Polymer Electrolyte Membrane (PEM)

The electrolyte membrane's appearance varies among manufacturers, with Nafion, produced by DuPont, being a common choice. This membrane resembles the plastic wrap used for food storage but is significantly thicker, ranging from 50 to 175 μm. For comparison, standard writing paper is about 25 μm thick, making these membranes comparable to the thickness of 2–7 sheets of paper stacked together.

In operation, the membrane is kept moist, resembling a damp, thick plastic wrap. These PEMs are robust and stable, effectively separating hydrogen fuel from oxidant air. Electrons generated on one side of the cell cannot pass through the membrane. Instead, they travel through an external circuit to the other side of the cell, completing the electrical circuit necessary to power devices like cars or power plants.

4.2.1.4 The Electrodes

Electrochemical cell reactions involve electron exchange at two electrodes:

- *Anode*: The anode is where oxidation (electron loss) occurs and is electrically negative in a FC. It is made of platinum particles evenly distributed on carbon particles, facilitating the oxidation process.
- *Cathode*: Similarly, the cathode consists of platinum particles uniformly supported on carbon particles. The platinum serves as a catalyst, enhancing the reduction process without being consumed. The cathode's porous nature allows oxygen to permeate, facilitating the reduction reaction.

4.2.1.5 Bipolar Plates

Each individual PEMFC typically generates about 0.7 V. For practical applications requiring higher voltage, multiple cells are connected in series (refer to Figs. 4.2 and 4.3). However, due to ohmic and other resistances in the circuit,the total

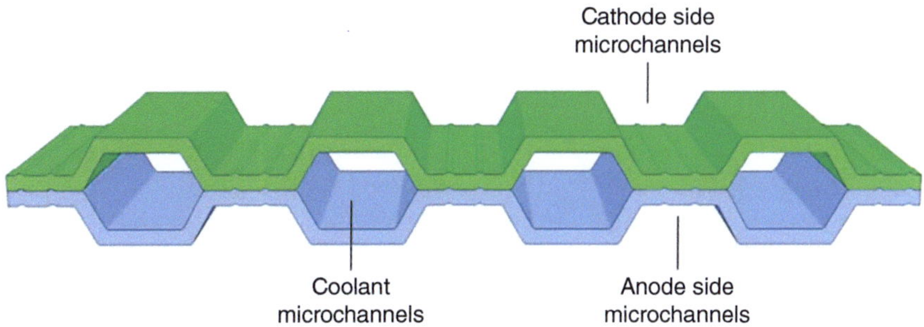

Fig. 4.2 Bipolar Plate Diagram. *Source:* National Renewable Energy Laboratory

voltage from directly connecting the cathode of one cell to the anode of the next is less than 1.4 V.

To efficiently connect numerous cells for higher voltage output, bipolar plates are used. These plates, made of conductive materials like carbon, stainless steel, titanium, or composites, have dual polarity. One side contacts the cathode, and the other contacts the anode. This design allows a single plate to separate cells while carrying hydrogen gas on one side and air on the other. The plates often feature corrugated flow channels for efficient gas distribution. The use of bipolar plates simplifies the stacking of cells, as shown in Fig. 4.2.

4.2.1.6 The PEMFC Stack

Many industrial applications, such as commercial electric motors, require voltages in the range of 200–300 V. To achieve this, individual FCs are connected in series to form a "stack."

In a stack, instead of having separate anode and cathode current collectors side by side, a single plate with flow fields on each side is used. This design minimizes the overall volume and weight of the stack and prevents the mixing of gases, which could lead to direct oxidation of the fuel. Bipolar plates also provide electrical conduction between cells and add structural strength to the stack.

Electrons produced at the anode on one side of a bipolar plate are conducted through the plate to the cathode on the other side. Two end-plates, one at each end of the stack, are connected via an external circuit.

The size of a single FC can range from a few square centimeters to a thousand square centimeters. A stack may consist of a few to hundreds of cells connected in series using bipolar plates. For example, in a passenger car, a stack might combine up to 400 FCs to produce up to 120 kW (163 hp) of direct current (DC) power output.

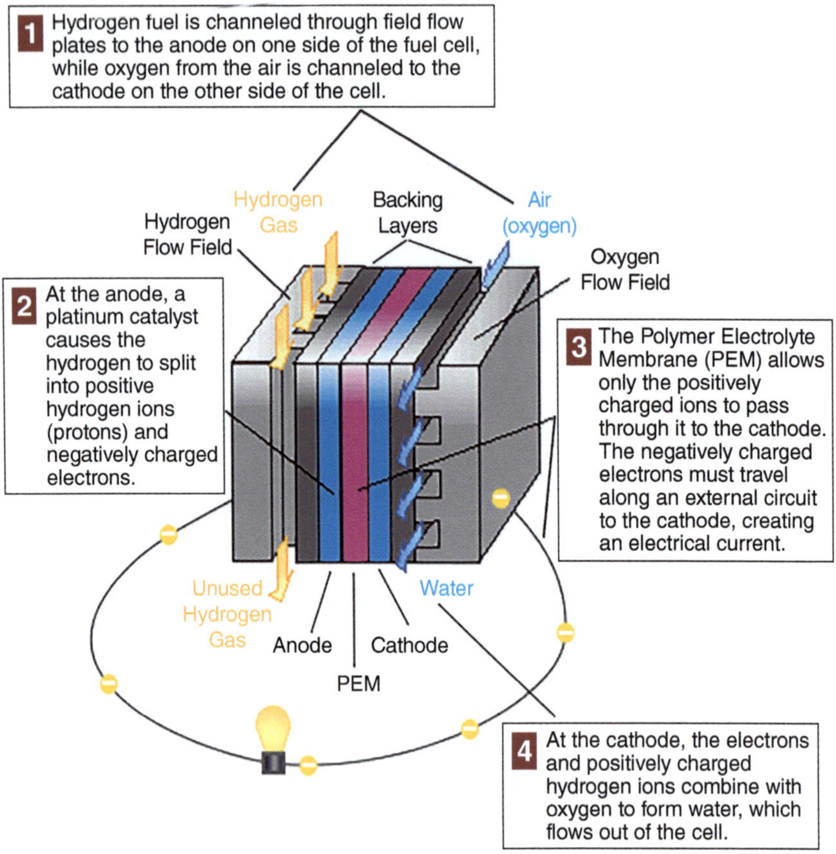

Fig. 4.3 PEMFC Stack Diagram. *Source:* https://www.fueleconomy.gov/feg/fcv_pem.shtml, as part of the DOE Office of Energy Efficiency & Renewable Energy

4.2.1.7 Electrode Poisoning

In PEMFCs, the electrodes are made with a platinum alloy catalyst at the anode, which is highly sensitive to carbon monoxide (CO) poisoning. This sensitivity can lead to a decrease in current output and a reduction in the FC's lifespan. The issue arises when the anode's catalyst sites are blocked by CO adsorption, preventing efficient FC operation. Various solutions have been explored to mitigate electrode poisoning, not only from CO but also from other species. For instance, alkaline FCs have been studied for carbon dioxide poisoning, which occurs due to the use of impure hydrogen or oxygen contaminated with carbon dioxide.

4.2.1.8 FC Reformers

FC reformers, or fuel processors, are devices designed to convert hydrogen-rich fuels into a form suitable for use by the FC. This reforming process involves the chemical decomposition of hydrocarbons or alcohols into hydrogen, generating heat and other by-products. It is crucial to remove these by-products to ensure the purity of the hydrogen gas. Depending on the fuel and the type of FC, the fuel processor may primarily focus on converting carbon monoxide to carbon dioxide, removing trace impurities like sulfur compounds in a sorbent bed, or employing a combination of reactors and sorbents.

Certain FCs, such as molten carbonate and solid oxide fuel cells (SOFCs), operate at high enough temperatures to allow for internal reforming, where the fuel reformation occurs within the FC itself. However, even with internal reforming, it is necessary to have traps to remove impurities from the unreformed fuel before it reaches the FC. Both internal and external reforming processes release carbon dioxide, but the quantities are significantly lower compared to the emissions from internal combustion engines typically used in transportation of goods and people.

4.2.2 Alkaline Fuel Cell (AFC)

AFCs, also known as Bacon fuel cells, are among the earliest developed FC technologies, with their initial development dating back to 1902. English engineer Francis Thomas Bacon made a significant advancement in 1932 by developing a 5 kW stationary AFC. These FCs gained practical application during the US Apollo space missions in the 1960s, where they were used for generating electrical energy and water onboard spacecraft. AFCs utilize a solution of potassium hydroxide in water as the electrolyte, which separates the anode and cathode. They can employ various nonprecious metals as catalysts for electrode production and operate within a temperature range of 80–260 °C. The concentration of potassium hydroxide varies from 35 to 85 wt%, with lower concentrations used in lower temperature AFCs.

A notable challenge for AFCs is their susceptibility to carbon dioxide (CO_2) poisoning, as even small amounts of CO_2 can significantly impact cell performance and durability due to carbonate formation. The use of recirculating liquid electrolytes can mitigate carbonate formation but introduces issues like shunt currents, increased corrosion, and stress on materials. Alkaline membrane fuel cells (AMFCs) have been developed to address these concerns, showing less vulnerability to CO_2 poisoning compared to liquid-electrolyte AFCs. Recent advancements in AFCs, particularly those using a polymer membrane as the electrolyte, have achieved efficiencies above 60% in space applications. AMFCs are considered for applications ranging from watts to kilowatts. However, they still lag behind PEMFCs in terms of performance due to issues like CO_2 exposure, membrane conductivity and durability, higher temperature operation, water management, power density, and anode electrocatalysis.

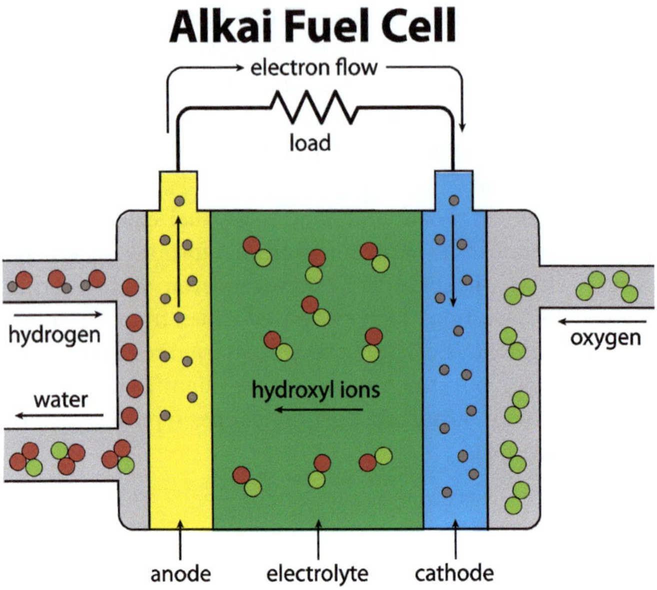

Fig. 4.4 Alkaline Fuel Cell (AFC) Diagram. *Source:* Adobe Stock; https://stock.adobe.com/images/fuel-cell-alkali-diagram/262639884

These FCs are similar to conventional PEMFCs but differ in the electrolyte medium used and the oxygen reduction product. The electrochemical reactions in AFCs are as follows:

$$\text{At anode: } H_2\left(g\right) \rightarrow 2H^{+}\left(aq\right) + 2e \tag{4.1}$$

$$\text{At cathode: } O_2\left(g\right) + 4e + 2H_2O\left(aq\right) \rightarrow 4\,OH^{-} \tag{4.2}$$

$$\text{Overall reaction: } H_2\left(g\right) + \tfrac{1}{2}O_2 \rightarrow H_2O\left(aq\right) \tag{4.3}$$

AFC technology has advanced significantly for practical use. Oxygen supply in the cathode compartment can be replaced by air, facilitated by soaking an anion exchange membrane with potassium hydroxide.

4.2.3 Electrodes in AFCs

The electrodes in AFCs are specifically designed to operate effectively in an alkaline environment.

Anode: Initially, nickel was the material of choice for anodes in AFCs. However, with advancements in materials science, metal hydride alloys like LaNi3 (AB3 type) have become preferred due to their superior performance.

These anodes are coated with an electrocatalyst layer, typically consisting of lanthanide–mischmetal-based Lm alloy [$LmNi_{4.1}Co_{0.4}Mn_{0.4}Al_{0.3}$]. The gas diffusion layer on both the anode and the cathode is created using poly(tetrafluoroethylene) (PTFE)-acetylene black on a nickel mesh. Common anode catalysts include nickel, platinum, and palladium. A recent innovation involves a mixture of various nitrates heated to form homogeneous samples of $Ni_{1-x}Co_xOy$ and YSZ (yttria-stabilized zirconia). This process produces crystallites of 30–40 nm, forming ceramics at high temperatures of 1,450 °C.

Cathode: In the initial stages of AFC development, platinum and other expensive metal catalysts were avoided, with nickel being the preferred choice due to its catalytic activity. Nickel electrodes were made from powdered nickel in two layers with different particle sizes to create a material with fine pores for gas diffusion. This material was suitable for both anode and cathode applications. Later, lithiated nickel (nickel plate treated with lithium metal) was used for cathodes. The Apollo mission FCs, for instance, utilized cathodes of this type.

For oxygen reduction in the alkaline medium at the cathode, MnO_2 emerged as an effective catalyst. A mixture of graphite and MnO_2, typically in a 1:1 ratio, is mechanically milled and combined with a polymeric binder like PTFE. A typical composition of graphite:MnO_2:PTFE in a 45:45:10 mass ratio has proven effective.

Siemens AFCs use silver as the cathode catalyst and Raney nickel for the anode, finding applications in environments like submarines. Generally, recommended cathode catalyst materials for AFCs include NiO, Au, Pt, or Ag, each offering unique benefits in terms of performance and cost.

4.2.4 Molten Carbonate Fuel Cell (MCFC)

MCFCs are emerging as a promising technology for power generation in electrical utilities, industrial settings, and military applications. These high-temperature FCs utilize an electrolyte composed of a molten carbonate salt mixture within a porous, chemically inert ceramic matrix made of lithium aluminum oxide. Operating at elevated temperatures around 650 °C (approximately 1,200 °F), MCFCs can use cost-effective, nonprecious metal catalysts at both the anode and the cathode.

One of the key advantages of MCFCs is their improved efficiency. When combined with a turbine, MCFCs can achieve efficiencies nearing 65%, significantly surpassing the 37–42% efficiencies typical of phosphoric acid FC plants. Additionally, when the waste heat from MCFCs is captured and utilized, the overall fuel efficiency can exceed 85%.

A unique feature of MCFCs, compared to other FC types like alkaline, phosphoric acid, and PEMFCs, is their ability to internally reform fuels. This means that fuels like natural gas and biogas can be converted into hydrogen directly within the FC through a process known as internal reforming, further reducing operational costs.

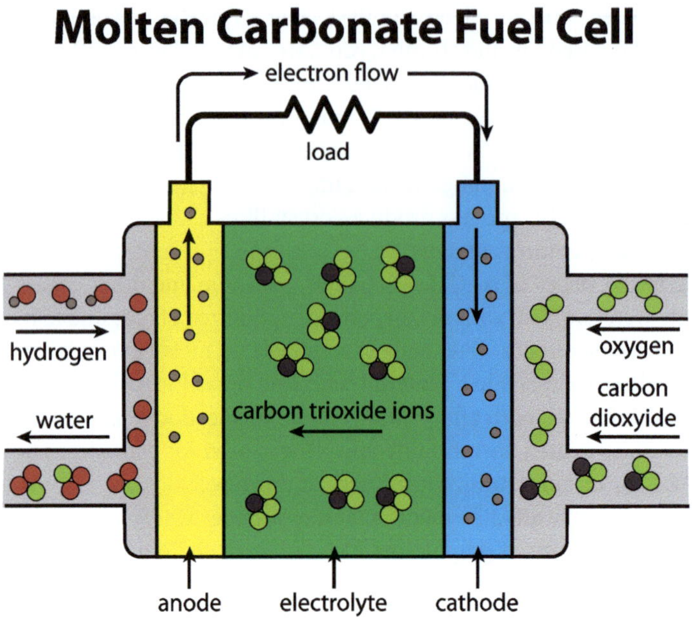

Fig. 4.5 Molten Carbonate Fuel Cell (MCFC) Diagram. *Source:* https://stock.adobe.com/images/fuel-cell-molten-carbonate-diagram/26263989

However, MCFC technology faces challenges in terms of durability. The combination of high operational temperatures and corrosive electrolytes accelerates wear and tear on components, leading to a shorter cell life. Advances in corrosion-resistant materials and innovative FC designs have improved durability, extending the operational life to around 40,000 hours.

Figure 4.5 illustrates an MCFC utilizing carbonate ions and hydrogen at the anode.

MCFCs operate effectively in the temperature range of 700–900 °C. In these cells, carbonate ions generated at the cathode migrate to the anode, facilitating the following electrochemical reactions:

$$\text{At anode: } H_2\,(g) + CO_3^{2-}\,(l) \rightarrow H_2O\,(g) + CO_2\,(g) + 2e + \text{heat} \tag{4.4}$$

$$\text{At cathode: } O_2\,(g) + CO_2\,(g) + 2e \rightarrow CO_3^{2-}\,(l) \tag{4.5}$$

$$\text{Overall reaction: } H_2\,(g) + \tfrac{1}{2}O_2\,(g) \rightarrow H_2O\,(g) \tag{4.6}$$

At temperatures between 600 and 700 °C, the alkali carbonates are molten, allowing the carbonate ion (CO_{32}^{-}) to be transported to the anode compartment through the matrix that separates the anode and the cathode. The subsequent sections will delve into the electrode materials and matrix used in MCFCs.

4.2.5 Electrodes in MCFCs

In the developmental stages of MCFCs, noble metals and metal oxides were the primary materials for electrodes. However, advancements in material science have led to the adoption of more cost-effective alternatives. Presently, MCFCs utilize anodes made of Ni-Cr/Ni-Al with a pore size of 3–6 μm and a porosity ranging from 45% to 70%. These anodes have a thickness between 0.20 and 1.5 mm. For the cathode, lithiated NiO is used, characterized by a pore size of 7–15 μm and a thickness of 0.5–1 mm.

MCFCs are known for their high fuel-to-electricity efficiency, nearing 60%, surpassing other lower-temperature FCs. Additionally, they can convert waste heat into electricity, pushing the overall thermal efficiency to approximately 85%.

4.2.6 Phosphoric Acid Fuel Cell (PAFC)

PAFCs utilize liquid phosphoric acid as the electrolyte, contained within a Teflon-bonded silicon carbide matrix. They feature porous carbon electrodes with a platinum catalyst. The electrochemical reactions in PAFCs are illustrated in the accompanying diagram.

The phosphoric acid electrolyte is absorbed by capillary action into the silicon carbide matrix embedded within the electrodes. PAFCs operate at around 200 °C. Regarded as the "first generation" of modern FCs, PAFCs are among the most mature and were the first to be commercially utilized. They are primarily used for stationary power generation, though some have powered large vehicles like city buses (Fig. 4.6).

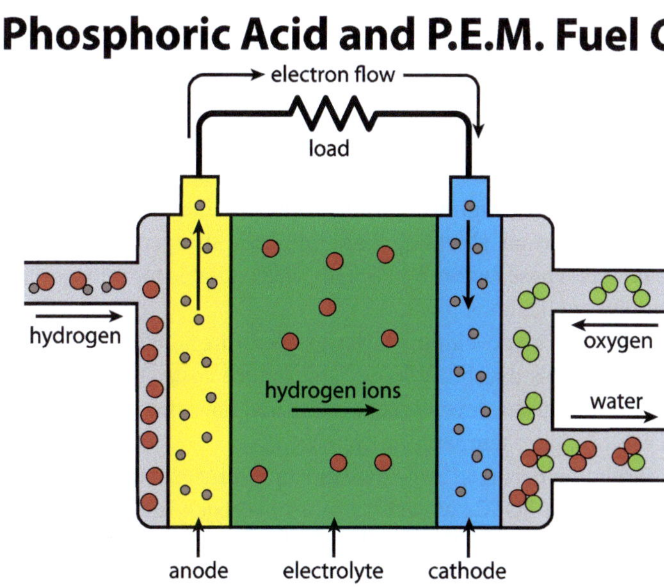

Fig. 4.6 Phosphoric Acid Fuel Cell (PAFC) Diagram. *Source:* https://stock.adobe.com/images/fuel-cell-phosphoric-diagram/262639899

PAFCs can tolerate impurities in reformed fossil fuels better than PEMFCs, which are susceptible to "poisoning" by carbon monoxide. In PEMFCs, carbon monoxide binds to the platinum catalyst at the anode, reducing efficiency. PAFCs are over 85% efficient in cogeneration of electricity and heat but less so in electricity generation alone (37–42%). Their efficiency is marginally higher than combustion-based power plants, which typically operate at around 33% efficiency. However, PAFCs are less powerful than other FCs for a given size and weight, leading to their typically large and heavy design. The high platinum catalyst loadings required for PAFCs also contribute to their higher cost.

4.2.7 Solid Oxide Fuel Cell (SOFC)

SOFCs utilize a hard, nonporous ceramic compound as their electrolyte. These cells are highly efficient, with about 60% efficiency in converting fuel to electricity. In co-generation systems that capture and use waste heat, the overall fuel use efficiency can exceed 85%.

SOFCs are available in three design configurations:

1) *Tubular Design*: The oxidant is inside the tube, fuel is on the outer surface, and the MEA is in between.
2) *Planar Design*: This involves bonding the MEA, flow channels, and gas collector plates to form a stack.
3) *Monolithic Design*: This uses a sintered, corrugated honeycomb structure of electrodes and electrolytes.

Operating at high temperatures (up to 1,000 °C or 1,830 °F), SOFCs eliminate the need for precious-metal catalysts, thus reducing costs. They can internally reform fuels, allowing for various fuel options and further cost reduction.

SOFCs are highly resistant to sulfur and are not poisoned by carbon monoxide, which can even be used as fuel. This makes them suitable for natural gas, biogas, and coal-derived gases. However, high-temperature operation leads to slow start-up, requires significant thermal shielding, and imposes stringent durability requirements on materials. These factors make SOFCs less suitable for transportation but viable for utility applications.

Current research is focused on developing lower-temperature SOFCs (operating at or below 700 °C) to address durability issues and reduce costs. These lower-temperature SOFCs are still in the development phase and have not yet achieved the performance levels of higher temperature systems.

In SOFCs, water formation occurs at the anode during hydrogen oxidation. The overall chemical reactions in SOFCs as well as other hydrogen-fueled cells are illustrated in Fig. 4.7. The anodic reaction involves hydrogen oxidation, while the cathodic reaction involves oxygen reduction.

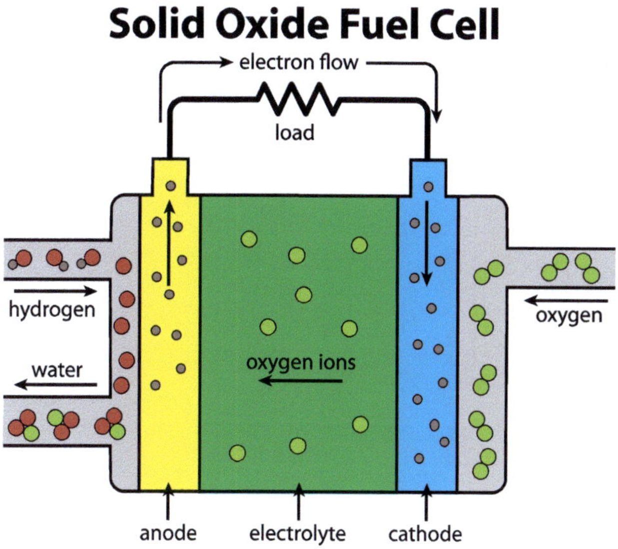

Fig. 4.7 Solid Oxide Fuel Cell (SOFC) Diagram. *Source:* https://stock.adobe.com/images/fuel-cell-solid-oxide-diagram/262639907

The reactions in SOFCs are as follows:

$$\text{At anode: } H_2\left(g\right) \rightarrow 2H^+ + 2e \tag{4.7}$$

$$2H^+ + O_2^- \rightarrow H_2O\left(g\right) \tag{4.8}$$

$$\text{At cathode: } ½O_2\left(g\right) + 2e \rightarrow O_2^- \tag{4.9}$$

At operating temperatures of 800–1,000 °C, water formed at the anode evaporates into the atmosphere. Oxygen ions (O_2^-) are transported from the cathode to facilitate the reaction at the anode.

In some SOFCs, hydrogen is mixed with carbon monoxide, leading to the exothermic reaction:

$$CO\left(g\right) + H_2O \rightarrow CO_2\left(g\right) + H_2\left(g\right) \tag{4.10}$$

This fuel mixture reduces the heating cost of the FC. SOFCs can also be designed in a cylindrical configuration, as depicted in Fig. 4.8. Stacking multiple SOFCs requires the use of interconnects, typically made of $LaCrO_3$ doped with rare earth elements like Sr, Ca, or Mg to improve conductivity. Yttrium chromite doped with Ca is another alternative, offering better thermal expansion properties than $LaCrO_3$.

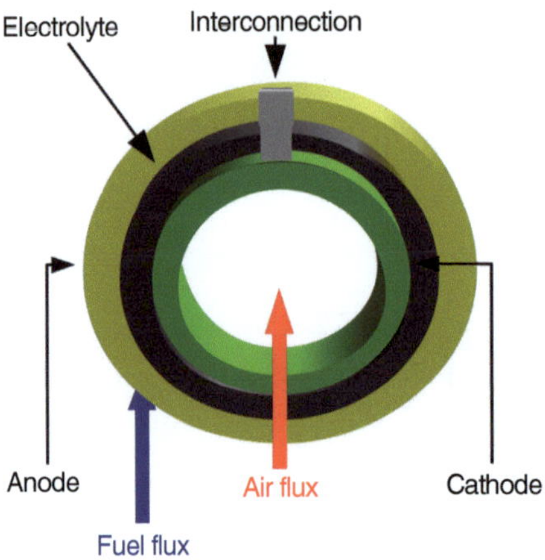

Fig. 4.8 Tubular Solid Oxide Fuel Cell (SOFC) Diagram. *Source:* https://commons.wikimedia.org/wiki/File:SOFC-en.svg

4.2.8 Electrodes in SOFCs

In SOFCs, the choice of materials for electrodes is crucial due to the high-temperature operation of the cell.

Anode: Initially, platinum was used for the anode in SOFCs, but due to its high cost, it has been replaced by more economical materials. The current anode is typically made from a composite of nickel and YSZ, which is permeable to hydrogen gas. This composite is preferred over pure forms due to differing thermal expansion rates, which can cause sintering and seal the necessary porosity for gas diffusion. The composite ensures thermal expansion compatibility, electrical conductivity, and permeability to hydrogen. It also withstands high temperatures without degradation in a hydrogen atmosphere.

Practical anode preparation involves applying a slurry of nickel, followed by electrochemical vapor deposition of YSZ. To maintain porosity, pore formers like thermosetting resins, starch, or carbon are used, followed by sintering. Alternative methods involve using nickel oxide-YSZ slurries and replacing traditional pore formers with a freeze-drying process to create more efficient pores.

Other materials, such as copper-cerium oxide, are being explored for anode use. Copper offers high electrical conductivity, while cerium oxide is effective for hydrocarbon oxidation.

Cathode: The cathode is typically a p-type perovskite, such as lanthanum manganate ($LaMnO_4$), doped with elements like Sr, Pr, or Ce to enhance conductivity. The cathode is a significant component, constituting about 90% of the FC's mass. Siemens Westinghouse's tubular SOFC design, for instance, features a 40-μm thick electrolyte layer applied to the cathode.

Electrolyte: The electrolyte in SOFCs must exhibit high ionic conductivity without electronic conductivity at operating temperatures. It should also be thermally, chemically, and structurally stable. Commonly used electrolyte materials include solutions like potassium hydroxide (KOH) in water, with an efficiency of about 70% and operating temperatures ranging from 85 to 200 °C. Yttria strontium-doped zirconia (YSZ) is another option for the electrolyte.

SOFCs operate at temperatures of 800–1,000 °C and are less sensitive to fuel purity compared to PEMFCs. They also do not face issues with water management and membrane stability. Siemens Westinghouse's pioneering work in SOFCs includes the tubular design shown in Fig. 4.8, which has been tested successfully for extended periods with minimal voltage loss. SOFCs are well suited for primary and auxiliary power sources in homes, office buildings, military installations, and other applications.

4.2.9 Direct Methanol Fuel Cell

Direct methanol fuel cells (DMFCs) differ from the majority of FCs, which are hydrogen-powered. DMFCs utilize pure methanol, typically diluted with water, and directly supply it to the FC's anode. This type of FC is essentially a power generator that converts methanol's chemical energy into electricity. Methanol as a fuel offers several advantages: it is lightweight, has a higher energy density than hydrogen, simplifies refueling, and eases storage issues commonly associated with hydrogen. Additionally, methanol's liquid state facilitates easier transportation and distribution using existing infrastructure. DMFCs are particularly effective for powering portable applications like cell phones and laptops (Fig. 4.9).

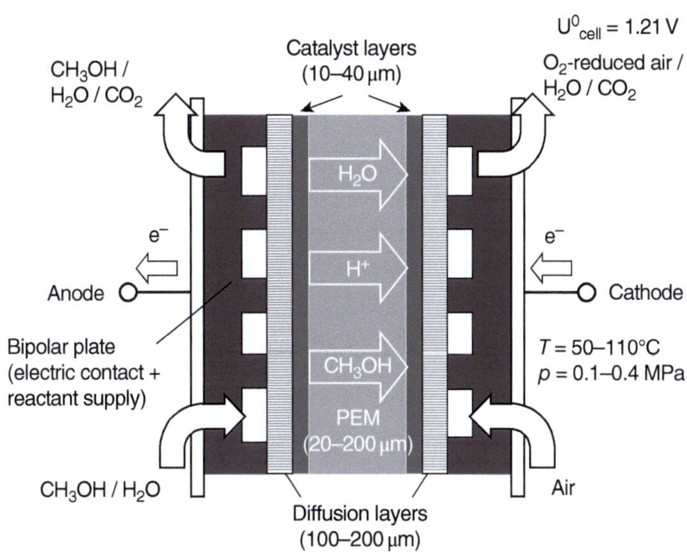

Fig. 4.9 Direct Methanol Fuel Cell (DMFC) Diagram. *Source:* https://www.sciencedirect.com/science/article/abs/pii/S0920586105001215

Table 4.2 Fuel Cell Operating Temperatures & Electrochemical Reactions

Fuel cell technology	Operating temperature (°C)	Electrochemical reactions
Polymer electrolyte membrane fuel cell (PEMFC)	60–100 °C	Anode: $H_2 \rightarrow 2H^+ + 2e^-$ Cathode: $½O_2 + 2H^+ + 2e^- \rightarrow H_2O$ Cell: $H_2 + ½O_2 \rightarrow H_2O$
Alkaline fuel cell (AFC)	90–100 °C	Anode: $H_2 + 2(OH)^- \rightarrow 2H_2O + 2e^-$ Cathode: $½O_2 + H_2O + 2e^- \rightarrow 2(OH)^-$ Cell: $H_2 + ½O_2 \rightarrow H_2O$
Phosphoric acid fuel cell (PAFC)	175–200 °C	Anode: $H_2 \rightarrow 2H^+ + 2e^-$ Cathode: $½O_2 + 2H^+ + 2e^- \rightarrow H_2O$ Cell: $H_2 + ½O_2 \rightarrow H_2O$
Molten carbonate fuel cell (MCFC)	600–1,000 °C	Anode: $H_2 + CO_3^{2-} \rightarrow H_2O + CO_2 + 2e^-$ Cathode: $½O_2 + CO_2 + 2e^- \rightarrow CO_3^{2-}$ Cell: $H_2 + ½O_2 + CO_2 \rightarrow H_2O + CO_2$ (CO_2 is consumed at the cathode and produced at the anode)
Solid oxide fuel cell (SOFC)	600–1,000	Anode: $H_2 + O^{2-} \rightarrow H_2O + 2e^-$ Cathode: $½O_2 + 2e^- \rightarrow O^{2-}$ Cell: $H_2 + ½O_2 \rightarrow H_2O$

Table 4.2 summarizes the operating temperature ranges and corresponding key electrochemical reactions for the fuel cell technologies discussed in this chapter.

4.3 FC Architecture

For the successful operation of automobiles or household systems, a significantly higher power output is necessary. This is achieved through a combination of FCs and operational convenience, embodied in the fuel cell power system (FCPS). The FCPS is an integration of various subsystems that collectively generate electrical energy in a user-friendly format. Its primary objective is to manage and process water, gases, and excess heat for efficient electrical energy generation.

4.3.1 FCPS Subsystems

A typical proton exchange membrane (PEM) FCPS includes several key subsystems:

1) *Fuel Cell Stack Subsystem (FCSS)*: This subsystem comprises the FC stack and its structural hardware. It is designed to maintain optimal operating conditions and interfaces with other subsystems and controls.

2) *Hydrogen Supply and Conditioning Subsystem (HS&CS)*: Responsible for converting and purifying hydrocarbon-based fuel into reformate, this subsystem also manages mass flow, pressure, and temperature.
3) *Oxidant Supply Subsystem (OSS)*: Typically includes an air compressor, filters, and devices to maintain required pressure, temperature, relative humidity, and mass flow control.

The FCPS is a sophisticated system that combines various technologies and components to efficiently generate and manage electrical energy. Its subsystems work in harmony to ensure optimal performance, making it a crucial component in modern FC applications. The integration of these systems demonstrates the potential of FCs as a sustainable and versatile energy source for the future.

To facilitate the cathodic reaction, FCs extract oxygen from incoming air. The efficiency of FCs increases with the pressure of reactant gases. Consequently, many systems incorporate an air compressor or blower to elevate the inlet air pressure to 2–4 times the ambient atmospheric pressure, typically around 3–4 bar. Some designs include an expander to reclaim energy from high-pressure exhaust gases. In transportation applications, air compressors should achieve at least 75% efficiency.

1) *Thermal and Water Management Subsystem (T&WMS)*: This subsystem is crucial for maintaining the FCPS's operating temperature, removing excess heat and water from the gas stream, and ensuring appropriate gas humidification. Since the PEM performs poorly when dry, many systems include a humidifier for the inlet air. These humidifiers often use a thin membrane, similar to the PEM material, to recycle water produced by the FC, thereby maintaining optimal hydration of the PEM.
2) *Controls and Communications Subsystem (C&CS)*: This subsystem combines hardware and software to manage all FCPS functions. It employs specific algorithms adaptable to various operating conditions, focusing on performance, durability, and efficiency optimization. Control functions are based on the interpretation of electrical data, including current, voltage, water conductivity, and material properties like pressure, temperature, gas consumption, and humidity.
3) *DC/DC or DC/AC Converter (Power Conditioner Subsystem)*: This subsystem interfaces between the FC stack output and the electrical parameters necessary for effective utilization of the generated electric power. It ensures compatibility with the requirements of automotive propulsion or residential use, which are typically 400 V DC and 110 V alternating current (AC). The conversion and conditioning processes only marginally reduce system efficiency, by approximately 2–6%.

4.3.2 Balance of Plant

All these subsystems are integrated into the FCPS, as illustrated in Figure 4.10. FCs produce electricity with low voltage and high DC output. For use in homes and workplaces, where AC is standard, this DC output must be converted and conditioned appropriately. This conversion is essential for compatibility with mobile transportation or utility power grids, ensuring that FC technology can be effectively utilized across various applications.

The balance of plant (BoP) is a crucial, power-intensive, and costly component of any FC system. It comprises a microprocessor that controls an array of subsystems based on thermodynamic principles, ensuring stable FC stack performance amid varying loads and environmental conditions. The BoP encompasses all direct stack support subsystems, including the reformer, compressors, pumps, and recuperating heat exchangers. It manages material flows (gases, coolant, and water) with varying heat contents and informational flows, facilitating interfaces between operators and machines through various sensors.

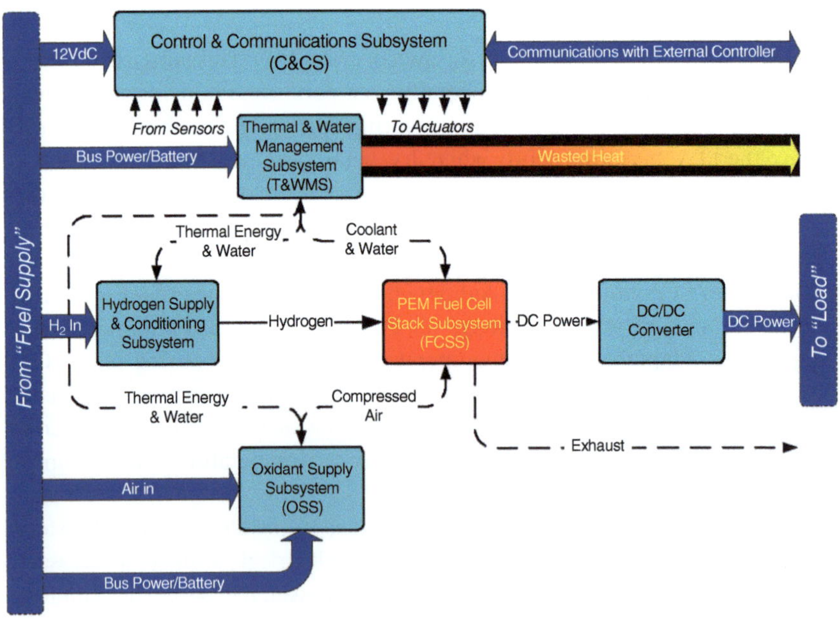

Fig. 4.10 Fuel Cell Power System for Transportation

4.3.3 Membraneless FC

Membraneless FCs represent an innovative approach to simplifying and reducing the cost of FC production. There are two primary principles of membraneless operation:

- *Laminar Flow Fuel Cells (LFFCs)*: These utilize nonmixing laminar flows, where the interface between the two flows acts as a proton/ion conductor. Leveraging microfluidics principles, LFFCs are designed with ultranarrow channels, preventing the mixing of solutions in the cathode and anode sections. While the absence of a membrane offers a cost-effective solution, their small size limits their application to portable devices requiring low power in the watts range.
- *Mixed Reactant Fuel Cell (MRFC)*: In contrast to LFFCs, MRFCs employ a mixture of fuel and electrolyte, relying on the electrodes' ability to separate oxidation and reduction reactions. MRFCs could potentially be simpler and more cost-effective than traditional FC systems, offering a promising alternative for various applications.

These advancements in membraneless FC technology reflect ongoing efforts to optimize FC efficiency and cost-effectiveness, potentially broadening their applicability in both portable and larger-scale energy solutions.

5

Hydrogen Technology Essentials

5.1 Hydrogen Safety

The introduction of hydrogen into the commercial marketplace necessitates critical safety considerations. Over the years, hydrogen has been perceived as a potentially hazardous substance, partly due to historical events like the 1937 Hindenburg dirigible fire and the development of the hydrogen bomb (H-bomb). However, it is important to note that the Hindenburg disaster was primarily caused by the flammable material of the airship's exterior, ignited by static electricity, not by hydrogen itself. Similarly, the nuclear fusion reactions in an H-bomb are unrelated to the chemical reactions of hydrogen used as a fuel in internal combustion engines (ICEs) or fuel cells. In reality, hydrogen has been safely utilized in various industries for years (Table 5.1).

Hydrogen is nontoxic and environmentally friendly. Unlike common fuels like oil and gasoline, which are toxic and can cause significant environmental damage, hydrogen does not produce harmful fumes or contribute to groundwater pollution. The only emissions from a hydrogen fuel cell are water and electricity. In the event of a hydrogen spill, its lightness causes it to evaporate quickly, leaving water as the only by-product. This contrasts with oil spills, which require extensive cleanup efforts and often result in long-term environmental damage.

Hydrogen's potential for explosiveness is a common concern. It becomes volatile in the atmosphere at a 4% concentration, which might seem alarming. However, compared to gasoline, which becomes volatile at very low percent concentration, hydrogen is relatively safer. Its small molecular size, high diffusivity, and buoyancy are advantageous safety properties. These characteristics enable hydrogen to disperse rapidly in the air, reducing the likelihood of combustible situations (Table 5.2).

In contrast, fossil fuels like propane or gasoline tend to accumulate near the ground, increasing the risk of ignition and explosion. For a hydrogen fire to occur,

Hydrogen Energetics: A Future Clean Energy Carrier, First Edition. Roman J. Press.

Table 5.1 Hydrogen Safety-Related Properties

Hydrogen does NOT	
• Explode in open air	• Endanger water
• Decompose	• Smell badly
• Self-ignite	• Cause cancer
• Oxidate	• Damage or risk pregnancies
• Corrode	• Cost prohibitively
• Produce radiation	• Innately become toxic

Table 5.2 Fire Hazards Characteristics

Property	Gasoline	Methane	Hydrogen
Density (kg/m^3)	4.40	0.65	0.084
Diffusion coefficient in air (cm^2/s)	0.05	0.16	0.610
Specific heat at constant pressure (J/Gk)	1.20	2.22	14.89
Ignition limits in air (vol%)	1.0–7.6	5.3–15.0	4.0–75.0
Ignition energy in air (Mj)	0.24	0.29	0.02
Ignition temperature (°C)	228–471	540	585
Flame temperature in air (°C)	2197	1875	2045
Explosion energy (G TNT/kJ)	0.25	0.19	0.17
Flame Emissivity (%)	34–43	25–33	17–25

Source: Nejat Veziroglu with permission of University of Miami.

specific conditions must be met, including a precise concentration of hydrogen, an ignition source, and the correct amount of oxidizer. This makes accidental hydrogen fires less likely compared to other fuels (see Fig. 5.1).

At lower concentrations, hydrogen requires a significant energy input to ignite, comparable to gasoline and natural gas within their respective flammability ranges. This characteristic makes hydrogen less prone to ignition at its lower flammability level.

In environments lacking oxidizers, hydrogen cannot explode in tanks or containers dedicated solely to hydrogen storage. This contrasts with gasoline, which can become volatile and potentially explosive with just 1–3% concentration of pure oxygen. Additionally, the likelihood of hydrogen exploding in open air is

Fig. 5.1 Gaseous Hydrogen Safety Sign. *Source:* middlenoodle/Adobe Stock Photos

diminished due to its rapid dispersal. In contrast, gasoline vapors, heavier than air, linger close to the ground, presenting a substantial risk of explosion.

Hydrogen fires, characterized by a nearly invisible, pale blue flame, burn intensely briefly due to the gas's high flame speed and rapid dissipation. This quick burnout, coupled with minimal heat radiation from hydrogen flames, reduces the risk of adjacent materials igniting and the fire spreading. For instance, a gasoline fire in a vehicle that might last around 30 minutes would be significantly shorter in a hydrogen-fueled scenario. This rapid burnout also makes hydrogen a potentially safer fuel for aircraft than current aviation fuels, especially in its liquid state.

5.1.1 Liquid Hydrogen Safety

As a liquid, hydrogen is colorless, and its vapors are colorless, odorless, tasteless, and highly flammable. While noncorrosive, liquid hydrogen's extreme cold necessitates specialized materials for handling and containment. Equipment must be designed for low-temperature operation, with vessels and piping chosen to withstand the associated pressures and temperatures, obeying relevant codes and regulations.

5.1.2 Flammability

Hydrogen's broad flammability range, spanning from 4% to 74% in air, combined with its low ignition energy threshold, demands meticulous handling to avoid accidental hydrogen–air mixtures. It is crucial to eliminate potential ignition sources, including sparks from electrical equipment, static electricity, open flames, or excessively hot objects. Mixtures of hydrogen and air within this flammable range are predisposed to explosion and typically burn with a pale blue flame that is nearly invisible (see Fig. 5.2).

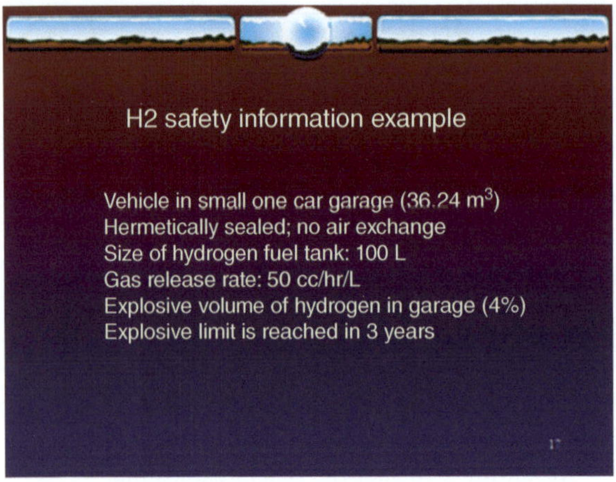

Fig. 5.2 Hydrogen Safety in the Garage Calculation

5.1.2.1 Containers

Liquid hydrogen is typically stored in on-site systems comprising a tank, vaporizer, and control mechanisms. These systems are tailored based on the rate of usage, pressure requirements, and regulatory standards.

5.1.2.2 Tanks

These are generally cylindrical and positioned horizontally, though vertical cylindrical and spherical tanks are also utilized. Standard tank sizes for industrial applications vary from 1,500 to 25,000 gal. Featuring vacuum insulation, these tanks are equipped with pressure relief valves designed according to ASME specifications for safety.

5.1.3 Transferring Liquid Hydrogen

It is a safety protocol to have two individuals present during the usage, transfer, or movement of liquid hydrogen containers. This guideline, however, does not apply to specially trained personnel from liquid hydrogen-supplying companies who are accustomed to handling this substance. Typically, hydrogen is vaporized for use as a gas. Extracting liquid hydrogen from a tanker, tank, or cylinder necessitates a closed system equipped with appropriate safety relief mechanisms. This system should be capable of being evacuated or purged to prevent the formation of flammable or explosive mixtures of liquid air and hydrogen. For purging, helium is recommended, as liquid hydrogen can solidify gases like nitrogen, potentially causing blockages or ruptures in the transfer line or storage vessel. Liquid transfer

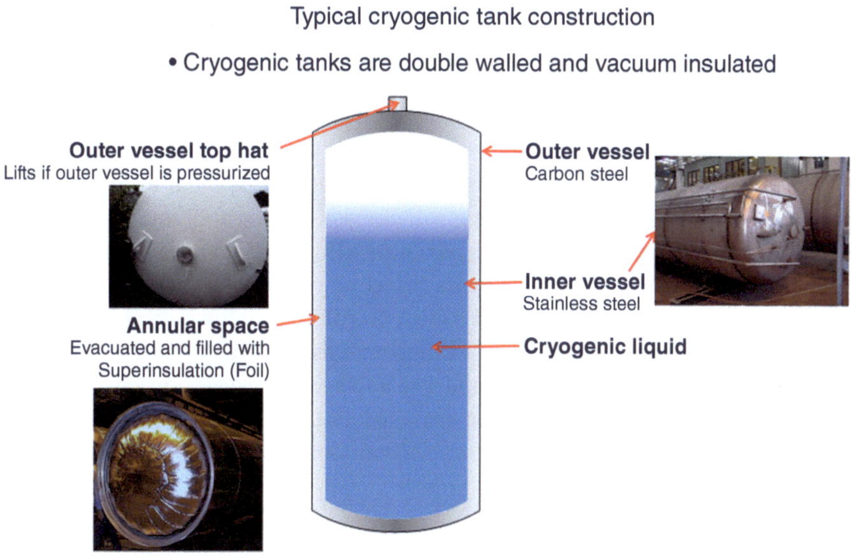

Fig. 5.3 Typical Liquid Hydrogen Tank. *Source:* US Department of Energy/Public Domain

lines must be vacuum insulated to reduce vaporization or formation of liquid air, which can lead to oxygen enrichment. All equipment involved in the transfer process must be electrically grounded and bonded (see Fig. 5.3).

5.1.3.1 Shipment

Liquid hydrogen is typically transported in semitrailers with capacities ranging from 12,000 to 17,000 gal. These tankers, similar in design to stationary tanks, also comply with the Department of Transportation's regulations. They are used to refill stationary tanks.

5.1.4 Safety Considerations

Handling liquid hydrogen involves risks such as fire, explosion, asphyxiation, and exposure to extremely low temperatures. For comprehensive safety guidelines on gases and equipment, consult the Air Products Safety Data Sheet (SDS). Hydrogen's potential to form and ignite flammable mixtures is notably higher than other gases due to the following reasons:

1) Hydrogen's ability to quickly seep through very small openings.
2) The extremely low ignition energy required for hydrogen mixtures. Hydrogen's flame is pale blue to almost invisible, increasing the risk of accidental exposure

to a hydrogen fire. To mitigate these risks, appropriate design and operational procedures are crucial. This includes preventing combustible fuel-oxidant mixtures and removing potential ignition sources like electric sparks, static electricity, and open flames.

3) Effective evacuation and purging are critical to avoid the formation of flammable or explosive mixtures. Adequate ventilation is essential to reduce the risk of flammable mixtures during a hydrogen leak or spill and to prevent asphyxiation. Protective clothing is recommended to safeguard against extreme cold from liquid hydrogen and vapors as well as potential flash fires.
4) Cold burns are a hazard due to brief contact with frosted lines, dripping liquid air from cold lines or vent stacks, vaporizer fins, and vapor leaks. At liquid hydrogen temperatures, air can condense and become oxygen enriched due to nitrogen vaporization. This oxygen-enriched air can significantly increase the combustion rate of flammable materials, including clothing.

The widespread public use of hydrogen as an energy carrier necessitates a thorough examination of its safety. To establish hydrogen as a viable commercial fuel, both actual and perceived safety concerns must be fully addressed. Hydrogen has been safely used in industrial settings for years, and its properties can offer safety advantages over traditional fuels like gasoline.

5.1.5 Hydrogen–Ammonia Blend Safety

Ammonia usage presents notable challenges, particularly concerning safety. As a toxic gas, ammonia has an Occupational Safety and Health Administration (OSHA) exposure limit of 50 parts per million (ppm). Its health effects are detailed in Table 5.3. Despite these risks, ammonia is a widely used chemical in industrial and agricultural settings, and it can be managed safely (Fig. 5.4).

Table 5.3 addresses the acute health effects of short-term exposure to ammonia. However, the impacts of prolonged, low-level exposure are less understood. Research involving rats exposed to 180 ppm continuously for 90 days showed no significant organ or tissue abnormalities.

Ammonia's current use is predominant in industrial and rural farming circumstances. Its broader application in densely populated urban areas raises questions about societal acceptance. Ammonia's odor is detectable at concentrations as low as 5 ppm. Therefore, minor leaks or spills at filling stations could incrementally increase airborne ammonia levels locally. In urban centers with heavy traffic, the use of ammonia for fuel storage might lead to noticeable odor levels, potentially conflicting with the image of a "clean" fuel.

The European Union Directive classifies ammonia as environmentally hazardous due to its high toxicity to aquatic life. A shift to ammonia as a mainstream fuel

Table 5.3 Acute Health Effects of Ammonia

Effect	NH_3 Concentration in Air (By Volume)
Least perceptible odor	5 ppm
Readily detectable odor	20–50 ppm
No discomfort or impairment of health for prolonged exposure	50–100 ppm
General discomfort and eye tearing; no lasting effect on short exposure	150–200 ppm
Severe irritation of eyes, ears, nose, and throat; no lasting effect on short exposure	400–700 ppm
Coughing, bronchial spasms	1,700 ppm
Dangerous, less than ½ hour exposure may be fatal	2,000–3,000 ppm
Serious edema, strangulation, asphyxia, rapidly fatal	5,000–10,000 ppm
Immediately fatal	10,000 ppm (1%)

Fig. 5.4 Ammonia Safety Sign. *Source:* middlenoodle/Adobe Stock Photos

could lead to spills affecting water bodies, including lakes, rivers, and coastal ocean areas (see Table 5.3).

Another concern is the diversion of ammonia for illicit purposes, such as the production of methamphetamine. This issue is particularly prevalent in the Midwest farming regions of the United States, where ammonia is readily available. The Minnesota Department of Agriculture has issued guidelines addressing health hazards and spill responses for first responders and ammonia users.

The potential use of ammonia for onboard storage or as a hydrogen distribution carrier raises numerous safety and liability issues. Despite these challenges, ammonia's established role in industrial and agricultural sectors demonstrates that it can be managed with appropriate safety measures.

When considering the integration of hydrogen into the US natural gas pipeline system, safety concerns are complex and location specific. Generalized safety assertions are challenging due to the multitude of factors involved. Currently, there is a lack of comprehensive statistical data on the safety implications of blending hydrogen with other chemical components.

5.1.6 Codes and Standards for Safety

Hydrogen has been safely utilized in various industries, including chemical, metallurgical, food, and space exploration. As hydrogen begins to play a crucial role in global energy solutions, ensuring safety in its use as a fuel is paramount. The rise in hydrogen-fueling stations, fuel cell applications in telecommunications, and other hydrogen-based projects necessitates stringent safety measures. Codes and standards for storage, electrolyzers, fuel cells, transportation, and infrastructure are vital to overcome commercialization barriers and enhance public understanding. The commercialization of hydrogen as an energy carrier should align with local, state, federal, and international standards and permits.

The National Renewable Energy Laboratory (NREL), in collaboration with the Department of Energy (DOE), has developed and validated the Hydrogen, Fuel Cells, and Infrastructure Technologies (HFCIT) Program. This program integrates building codes and equipment standards for hydrogen systems across commercial, residential, and transportation sectors, providing technical resources for development. A comprehensive list of key US and international organizations developing hydrogen codes and standards is available at NREL's website.

5.1.7 Hydrogen Sensors

Effective hydrogen safety relies on the implementation of hydrogen sensors. These sensors convert detected gas concentrations into electrical signals for measurement and safety analysis. Based on various principles, including electrochemical and optical, current research focuses on range, sensitivity, selectivity, and response time.

NREL is evaluating hydrogen sensors for leak detection and gas purity monitoring. These sensors are crucial for the safety of hydrogen-fueling stations and related equipment. NREL is also investigating polymer films for coating pipes and joints, Teflon composite guttering gasket materials, and integrating sensor chemicals into coatings and gaskets. The lifespan of hydrogen sensors varies based on their operating principles. For example, electrochemical sensors for gases like carbon monoxide or hydrogen sulfide typically have a lifespan of two to three years.

5.1.8 Hydrogen Safety-Related Properties

Hydrogen does not create fumes or contribute to groundwater pollution. The only emissions from a hydrogen fuel cell are water and electricity. In contrast, common fuels like oil and gasoline are toxic and environmentally harmful. A hydrogen spill results in evaporation, leaving water as the only by-product. Despite its lightweight, hydrogen has the highest energy-to-weight ratio of all known fuels.

Table 5.1 in the beginning of this chapter outlines the characteristics of gasoline, methane, and hydrogen regarding ignition and explosion hazards. Hydrogen's broader ignition range raises concerns about increased ignition probability compared to natural gas. Risk assessments for hydrogen blends in natural gas pipelines vary based on concentration, pipeline type, material, and age. Large-scale energy systems, including nuclear, fossil fuel, and renewable energy systems, present distinct risks. Natural gas systems generally pose lower severe accident risks than coal, petroleum, nuclear, and hydropower systems, but higher risks than wind and solar energy.

Risk management practices, such as installing monitoring devices, can mitigate risks in distribution mains and service lines. Introducing more than 50% hydrogen significantly increases the overall risk. These findings are specific to the US natural gas supply system and do not apply to new, dedicated hydrogen pipelines.

Understanding the behavior of hydrogen in various environments is crucial for developing effective safety protocols. Research into hydrogen's interaction with different materials, its behavior under varying pressure and temperature conditions, and its response to containment failures is ongoing. This research is essential for developing robust safety measures in hydrogen storage, transportation, and usage.

Furthermore, public education and awareness are key components in ensuring hydrogen safety. Dispelling myths and misconceptions about hydrogen, while providing accurate information on its properties and safe handling practices, will play a significant role in its acceptance as a mainstream energy source.

The DOE report concludes that hydrogen, when used as a motor vehicle fuel, does not present unusual safety risks. The report emphasizes that a well-designed hydrogen-powered vehicle and refueling infrastructure would likely pose no greater risk than conventional motor vehicle fuels.

5.2 Energy Storage Technologies

Energy storage encompasses various methods, including:

- *Thermal Storage*: Utilizing heat and cold to generate energy on demand or to reduce energy requirements.
- *Mechanical Storage*: Harnessing kinetic or gravitational forces through innovative technologies.

- *Chemical Storage*: Primarily involves converting electricity into gas or utilizing chemical reactions to create energy storage devices.
- *Pumped Hydropower*: Generating large-scale energy reservoirs using water.
- *Hydrogen Storage*: Converting surplus electricity into hydrogen, predominantly through electrolysis.

Short-term energy storage systems like supercapacitors, superconductive magnetic coils, or flywheels are not suited for hydrogen energy storage. Similarly, medium-duration storage systems such as pumped hydropower, compressed air energy storage, thermoelectric storages, and various electrochemical batteries (like lithium-ion, lead-acid, high-temperature, and flow batteries) are ineffective for hydrogen storage.Long-term storage methods, capable of efficiently storing and releasing vast amounts of energy over weeks to months, include pumped-storage hydroelectric (PHS) dams, rechargeable batteries, thermal storage including molten salts, and cryogenic systems. Our focus is on high-capacity storage options relevant to the hydrogen economy.

5.2.1 Chemical Storage

5.2.1.1 Flow Batteries

Flow batteries, also known as regenerative fuel cells, store and release energy through a reversible electrochemical reaction between two redox solutions. While promising for long-duration storage, they have not gained widespread market popularity. Vanadium redox-flow batteries are commercially available, but their adoption is hindered by high electrolyte costs and maintenance complexities.

Electrolyte options include zinc bromide (ZnBr), vanadium bromide (VBr), and sodium bromide (NaBr). These batteries function by circulating a solution across a membrane, facilitating ion exchange to charge or discharge the cell. The cell voltage, typically ranging from 1.0 to 2.43 V, is determined by the Nernst equation. Currently, there is limited technical and economic data available for their use in storing renewable energy. Commercial applications are mostly confined to long-duration backup power for the grid. The storage capacity of a flow battery is directly proportional to the volume of the electrolyte solution, making it analogous to both a fuel cell and an electrochemical accumulator cell.

5.2.1.2 Powerpaste

Powerpaste, a fluid gel developed by the Fraunhofer Institute for Manufacturing Technology and Advanced Materials (IFAM), represents a significant innovation in hydrogen storage. Created by reacting magnesium powder with hydrogen to form magnesium hydride at 350 °C and five to six times atmospheric pressure, Powerpaste releases hydrogen when it reacts with water. This reaction, described by the formula

$$MgH_2 + 2H_2O \rightarrow 2H_2 + Mg\left(OH\right)_2$$

indicates that half of the hydrogen is derived from the Powerpaste itself, and the other half from water.

Fraunhofer's research suggests that Powerpaste can store hydrogen energy at a density 10 times greater than a lithium battery of comparable size, making it a promising candidate for automotive applications. This development could revolutionize hydrogen storage, offering a safe, efficient, and high-density energy solution. Powerpaste delivers impressive energy densities of approximately 1.6 kWh/kg and 1.9 kWh/L at the material level.

The innovative aspect of Powerpaste lies in its ability to release hydrogen on-demand, making it an ideal candidate for applications where space and weight are critical factors, such as in electric vehicles (EVs) or portable power sources. Its high energy density and stability at ambient temperatures also make it a viable alternative for long-term energy storage, potentially transforming the way we store and use renewable energy.

5.2.1.3 Power to Gas

Power to Gas (P2G) technology involves converting electricity, primarily through electrolysis, into gaseous fuels like hydrogen or methane. Methane, the simplest hydrocarbon with the molecular formula CH_4, presents a manageable alternative to hydrogen in terms of storage and transportation. Its existing infrastructure, including pipelines and power plants, is well established and efficient.

Synthetic natural gas (SNG), also known as syngas, is produced through a multistep process beginning with hydrogen and oxygen. The Sabatier process then combines hydrogen with carbon dioxide to yield methane and water. This methane can be stored for later electricity generation, with the added benefit of water recycling to minimize resource consumption. During electrolysis, oxygen is stored for methane combustion in a pure oxygen environment, significantly reducing nitrogen oxide emissions. Methane combustion generates carbon dioxide and water. The carbon dioxide can be recycled to enhance the Sabatier process, while the water is reused in electrolysis, creating a sustainable cycle.

P2G systems can be single-stage, directly producing methane through technologies like reversible solid oxide cells (RSOCs), or two-stage, where hydrogen is further processed into syngas, methane, or LPG.

P2G's versatility allows the gas produced to serve as a chemical feedstock or be reconverted into electricity using standard generators, such as gas turbines. This method of storing and transporting energy in gas form, often through existing natural gas infrastructure, positions P2G as a leading solution for seasonal renewable energy storage.

5.2.1.4 Power to Liquid

Power to liquid (P2L) technology, akin to power to gas, involves converting hydrogen into liquid fuels such as methanol or ammonia. These liquids are easier to handle and transport than gaseous hydrogen and require fewer safety measures. P2L fuels are versatile, serving as electrochemical storage mediums for various transportation modes, including aircraft, and have applications in both industrial and power sectors.

5.2.1.5 Pumped-hydro Storage

Pumped-hydro storage (PHS), a significant renewable energy source, is predominantly utilized in hydroelectric dams. These dams can store substantial water volumes, effectively averaging the annual river flow between dry and wet seasons. While they do not directly store energy from intermittent sources like solar or wind, they play a crucial role in grid balancing by conserving water when alternative energy sources are active.

Energy storage in hydroelectric systems is primarily achieved through PHS. This method involves elevating water to higher reservoirs during low-demand periods using electric pumps. When energy is needed, water is released back through turbines, converting gravitational potential into electricity. Countries like Norway, Wales, Japan, and the United States have harnessed this technique, utilizing natural topography for reservoir creation. Norway, for example, with its vast hydroelectric capacity, has significant potential for expansion in this area.

Additionally, surplus power can be transformed into methane via the Sabatier process and stored in the natural gas network. Pumped-hydro facilities typically consist of two reservoirs at different elevations. Energy is stored by pumping water to the upper reservoir and released to generate electricity as it flows back down. This method is primarily used in large-scale energy plants and requires significant space.

Hydroelectric dams with reservoirs also contribute to energy storage. They release stored water to generate electricity during peak demand, functioning similarly to pumped storage but without the energy loss due to pumping. This method is highly efficient, as it only alters the generation timing, with hydroelectric turbines having a start-up time of just a few minutes.

The Llyn Stwlan dam of the Ffestiniog Pumped Storage Scheme in Wales exemplifies artificial energy storage and conversion, with a capacity to generate 360 MW of electricity for several hours (Fig. 5.5).

Fig. 5.5 Llyn Stwlan Dam. *Source:* Arpingstone/Wikimedia Commons/Public Domain

Bulk energy storage is largely dominated by hydroelectric dams, both conventional and pumped. These facilities store gravitational potential energy, which can be converted to electricity on demand.

The Sir Adam Beck Generating Complex at Niagara Falls, Canada, includes a large PHS reservoir, providing an additional 174 MW of electricity during peak demand periods (Fig. 5.6).

Globally, PHS stands as the most extensive active grid energy storage system. As of March 2012, the Electric Power Research Institute (EPRI) noted that PHS comprises over 99% of the world's bulk storage capacity, equating to approximately 127,000 MW. The efficiency of PHS typically ranges between 70% and 80%, with some instances reaching up to 87%.

PHS operates by utilizing excess electrical generation capacity during low-demand periods to pump water from a lower elevation to a higher reservoir. When electricity demand increases, the water is released back to a lower elevation, passing through turbines to generate electricity. Most PHS systems use the elevation difference between two water bodies, with reversible turbine-generator assemblies functioning as both pumps and turbines. Pure pumped-storage plants solely transfer water between reservoirs, whereas "pump-back" systems combine pumped storage with traditional hydroelectric methods that utilize natural stream flow.

5.2.1.6 Underground Hydrogen Storage

Underground hydrogen storage (UHS) involves storing hydrogen in geological structures such as caverns, salt domes, and depleted oil and gas fields. This

Fig. 5.6 Adam Beck Generating Station, Niagara Falls, Canada. *Source:* Ontario Power Generation/Wikimedia Commons/CC BY 2.0

method has been used for storing large quantities CO_2, town gas storage, methane storage, and recently for hydrogen storage. Worldwide underground structures operating as hydrogen storage are very limited and geological, engineering, economic interrogations including monitoring microbiological and geochemical activities require more data collection. The underground storage of pure hydrogen or the gas blend contained hydrogen is an interesting possibility for hydrogen energetics future.

5.2.1.7 Aluminum as an Energy Source

Aluminum, known for its high electrochemical potential (8.04 Ah/cm^3 compared to lithium's 2.06 Ah/cm^3), is explored as an energy store or hydrogen source. To extract energy, aluminum reacts with water to produce hydrogen. This process necessitates removing aluminum's natural oxide layer, which can be achieved through pulverization, chemical reactions with caustic catalysts like sodium hydroxide or alloys with gallium, mercury, and other metals substances. The reaction's by-product, aluminum oxide, can be recycled back into aluminum using the Hall–Héroult process, rendering the cycle potentially renewable. If powered by renewable sources like solar or wind, aluminum could serve as a highly efficient energy storage medium. However, systems utilizing this method are more complex than electrolysis and have not been commercially developed.

5.2.1.8 Home Energy Storage

The rise of distributed renewable energy generation is making home energy storage increasingly relevant. For homes equipped with photovoltaic systems, energy storage is crucial to achieve self-sufficiency beyond 40%. Currently, Li-ion batteries are preferred over lead-acid for home energy storage due to their superior performance. These systems can store surplus energy from home solar or wind sources, and the stored electricity can be sold back to the grid. Additionally, thermal energy storage, though less flexible, offers a cost-effective alternative to batteries. For instance, a standard 52-gal electric water heater can store about 12 kWh of energy, useful for heating purposes.

5.2.1.9 Grid Electricity and Power Stations

Renewable energy integration into the grid often involves thermal storage, which temporarily stores or removes heat. Facilities like the 280 MW Solana Generating Station use salt tanks for efficient thermal energy storage, allowing electricity generation even after sunset. This setup enables the plant to produce around 38% of its rated capacity annually.

Innovative approaches like the Bonneville Power Administration's experimental program in the Northwestern United States utilize excess wind and hydro power. This system charges home appliances like space heaters and water heaters with surplus energy, which is later used for heating. Such systems demonstrate the potential of integrating renewable energy into the grid effectively.

Advanced methods like those used in the Solar Two project in the United States and the Solar Tres Power Tower in Spain employ molten salt to store and convert solar thermal energy into electricity. This system circulates molten salt through structures to be heated by the sun, storing the heated solution in insulated tanks. Electricity is then generated by converting water to steam, which drives turbines.

5.2.1.10 Vehicle-to-Grid Storage

Vehicle-to-grid (V2G) storage is a technology that enables energy to be pushed back to the power grid from the battery of an EV. With V2G technology, an EV battery can be discharged based on different applications – such as energy production or energy consumption. This concept highlights the potential of EVs as mobile energy storage units, storing renewable energy within the global EV fleet contributing to grid stability and energy efficiency.

V2G technology can be used for bidirectional charging, which makes it possible to charge the EV battery and take the energy stored in the car's battery synonymously.

5.2.1.11 Economics of Energy Storage

The economic viability of energy storage is closely tied to the specific reserve service required. Various factors, including market uncertainties, significantly influence the profitability of different energy storage methods. Consequently, not all storage technologies are technically or economically feasible for storing multiple megawatt-hours (MWh) of energy. The optimal size and type of energy storage system are highly dependent on market conditions and geographic location.

The hydrogen storage is always combined with energy losses during hydrogen production or renewable energy processing, conversion to electricity, and energy transportation cost.

5.3 Hydrogen Storage for Transportation

Electric power's primary limitation is grounded in the inability to store electricity economically and directly. However, it can be stored in other forms, combined with significant energy consumption for conversion and distribution. Energy storage, particularly during off-peak times for use during peak demand, transforms erratic power into a reliable, high-value resource. This becomes increasingly important as fuel cell transportation and power generation gain traction, making hydrogen storage a pivotal aspect.

Hydrogen, despite having one of the highest specific energies, suffers from a low density, resulting in one of the lowest energy densities. For instance, 90 g of hydrogen occupies a volume of 1 m^3 (Fig. 5.7). Consequently, the chemical and

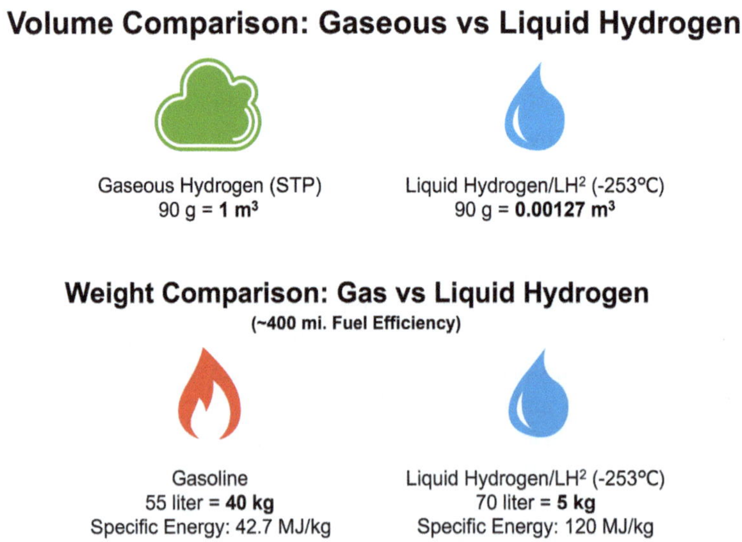

Fig. 5.7 Liquid Hydrogen Volume and Weight Comparisons

petroleum industries typically utilize hydrogen near its production site, transporting it in compressed or liquefied form due to its low energy density. The current hydrogen distribution infrastructure is inadequate for the expansive needs of a hydrogen-based economy.

A critical question is whether the existing US pipeline infrastructure, used for gasoline, natural gas, or propane, can adapt to transport compressed hydrogen and what alternative methods might be viable.

An alternative approach involves developing hydrogen-fueling stations at production sites. Storage options include compressed gas in pressure tanks, underground cavities, and super-insulated tanks for liquid hydrogen. Besides these traditional methods, other innovative storage forms are explored. Local production and storage offer the flexibility to harness various energy sources like solar, wind, biological, nuclear, and fossil fuel reforming. This approach necessitates further distribution via tanker trucks, with fueling stations supplying hydrogen for direct use in vehicle tanks.

For hydrogen-powered vehicles to be competitive, they must achieve a driving range exceeding 300 miles, requiring the storage of approximately 5 kg of hydrogen onboard. This presents significant scientific and engineering challenges, particularly in developing lightweight storage systems suitable for automotive applications.

For transportation needs, there are three primary hydrogen storage methods:

- *Compression and Storage in Pressure Tanks*: This method involves compressing hydrogen gas to high pressures, typically up to 700 bar, and storing it in specially designed high-pressure tanks. These tanks are made from advanced materials like carbon fiber to handle extreme pressures safely.
- *Liquefaction and Storage in Insulated Tanks*: Hydrogen can be cooled to its liquid state at −253 °C and stored in cryogenic tanks. This method significantly reduces the volume of hydrogen, making it more space-efficient. However, it requires substantial energy for cooling and sophisticated insulation to maintain the low temperatures.
- *Chemical Storage in Solid Compounds*: Hydrogen can be stored in solid-state materials through absorption or adsorption processes. Metal hydrides, chemical hydrides, and complex hydrides are some materials used for solid-state hydrogen storage. These materials can store hydrogen at lower pressures and moderate temperatures, offering a safer and potentially more compact alternative to gas or liquid storage.

Each of these storage methods has its advantages and challenges, and ongoing research aims to optimize them for efficiency, safety, and cost-effectiveness in various applications.

5.3.1 Storage of Hydrogen as a Compressed Gas

Compressed gas storage of hydrogen is a well-established method and is widely used for transporting small quantities of gas. This approach is prevalent across thousands of industrial and research settings, often employed for storing hydrogen produced by electrolyzes. Figure 5.8 illustrates hydrogen lightweight polymer tanks commonly used in this context.

In vehicular applications, these high-pressure tanks are typically mounted on the roofs of buses or integrated within the body of passenger cars. A standard high-pressure compressed gas system comprises a cylindrical tank, a pressure regulator to modulate hydrogen pressure for delivery to the propulsion system (fuel cell or ICE), gas flow control valves, tubing, mounting brackets, and environmental protection elements. Figure 5.8 depicts a representative high-pressure hydrogen storage system.

For remote and transportation applications, the emphasis is on low-cost, lightweight container manufacturing. The choice of material for the pressure vessel is crucial, given hydrogen's small molecular size and high diffusivity. Hydrogen can permeate materials impermeable to other gases and react with carbon in steel, leading to hydrogen-induced cracking. This issue is mitigated by using chromium-rich steels, alloys, or more commonly, composite-reinforced plastic materials. Modern high-pressure hydrogen storage systems typically operate at 350 bar (5,000 psi), with advanced carbon composite tanks capable of withstanding up to 700 bar (10,000 psi) within a temperature range of −30 to +50 °C.

Fig. 5.8 High-Pressure Polymer Tanks Storage

The three primary types of tanks include

- *Steel Tanks*: Traditional and sturdy, offering reliable containment.
- *Aluminum Core Encased in Composite Materials*: Utilizing fiberglass, carbon, or aramid fibers for reinforcement.
- *Plastic Core Encased in Composite Materials*: Similar to aluminum core tanks but with a plastic base, reinforced with fiberglass, carbon, or aramid fibers. In these composites, the fiber is impregnated with resin to create a continuous matrix.

The hydrogen lightweight polymer tanks installed in hydrogen bus are shown in Figure 5.9.

The tanks with compressed hydrogen are supposed to be held in the vehicle, typically on the roof space of buses or inside the passenger's cars body.

A high-pressure compressed gas system consists of a cylindrical tank, a pressure regulator that reduces the pressure of the compressed hydrogen to a lower value for hydrogen delivery to the propulsion system (fuel cell or ICE), gas flow control valves, tubing, mounting brackets, and some environmental protection.

While high-pressure storage poses certain safety challenges, hydrogen, when handled correctly, is no more hazardous than other commonly used flammable gases. Overall, the primary advantages of compressed gas storage are its simplicity, virtually unlimited storage duration, and a high number of possible refueling cycles.

Fig. 5.9 The Hydrogen LightWeight Polymer Tanks Installed in Hydrogen Bus. *Source:* scharfsinn86/Adobe Stock Photos

5.3.2 Storage of Hydrogen as a Liquid

Hydrogen can be effectively stored as a liquid, commonly referred to as LH_2, at temperatures of 20°K (−253 °C) and under a few bars of pressure in super-insulated tanks. LH_2 has been produced and distributed for various industrial applications for over 70 years. It is particularly useful for long-distance transportation, spacecrafts, and airplanes (Fig. 5.10).

The liquid hydrogen storage vessel depicted in Fig. 5.11 has supported space shuttle launches and other vehicles for over 30 years. Storing LH_2 necessitates specialized equipment. Containers are typically cylindrical and constructed from thin-wall metal alloys. Minimizing heat transfer from the external environment is crucial. This is achieved by surrounding the tanks with a thermal insulation barrier, consisting of a larger diameter cylinder concentric with the liquid container, separated by high vacuum and thermal insulation layers. Multilayer insulation includes several layers of metallic foil with thin glass wool between each layer.

The design and dimensions of an LH_2 (Fig. 5.11) storage system vary based on its intended use, whether in airplanes, buses, trucks, or cars. For example, the Messer Liquid Hydrogen storage system for an LH_2 vehicle has the following characteristics:

- Hydrogen content: 5 kg
- Temperature: 20°K

Fig. 5.10 Space NASA's Center, Liquid Hydrogen Tank. *Source:* Kim Shiflett/NASA/ Public Domain

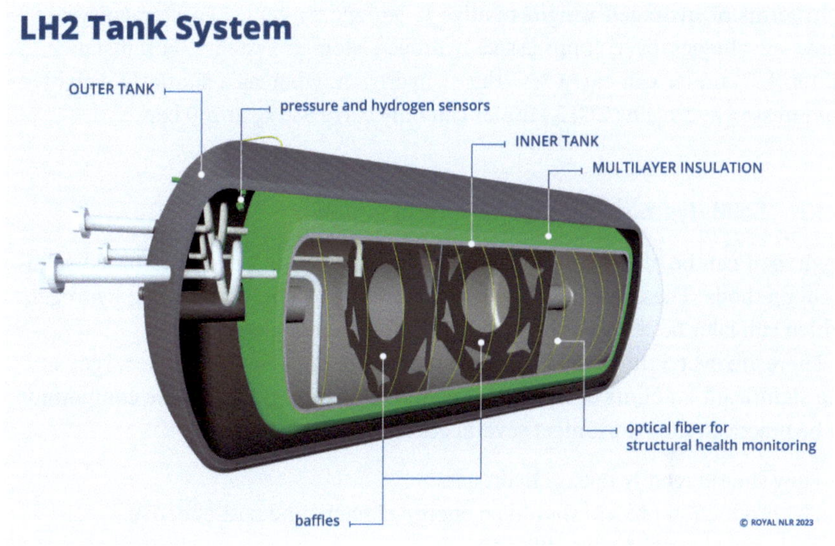

Fig. 5.11 Liquid Hydrogen Storage System. *Source:* Royal NLR, https://www.nlr.org/case/case-liquid-hydrogen-composite-tanks-for-civil-aviation/

- Pressure: 6 bar
- Gross volume: 130 L
- Net volume (LH_2): 71 L
- Tank mass: 50 kg
- Auxiliary mass: 40 kg
- Coolant flow rate: 100 L/h

LH_2-fueling stations, connected to storage vessels, are equipped with control systems and dispensers for transferring fuel to vehicle tanks. Refueling a 100 L LH_2 vehicle tank system typically takes less than two minutes.

Challenges with LH_2 storage:

1) The cooling process consumes significant energy, approximately 30–40% of the fuel's energy content, necessitating the development of more efficient cooling methods.
2) Tank manufacturing is costly.
3) Boil-off gas leads to hydrogen release and energy loss in the tank, approximately 4% per day.
4) Infrastructure investment is substantial, requiring special transfer lines with adequate insulation.
5) Filling and withdrawal processes demand stringent safety measures to prevent explosive mixtures.

In terms of hydrogen weight relative to storage content, LH_2 storage systems show advantages over compressed hydrogen storage systems. For instance, a 40-ton LH_2 trailer can carry 3,370 kg of hydrogen, whereas a similarly weighted compressed hydrogen (CGH_2) trailer can only carry 530 kg at 200 bar.

5.3.3 Solid Hydrogen Storage: Chemical Methods

Hydrogen can be efficiently transported and stored using chemical-based solid-state methods. These methods involve materials capable of absorbing hydrogen, which can later be released through heating or chemical reactions.

There are over a thousand chemical compounds capable (see Table 5.4) of storing significant amounts of hydrogen relative to their mass. For these compounds to be practical, they must meet several criteria:

- They should readily release hydrogen.
- The production process should be energy-efficient and cost-effective.
- They must be safe for handling and use.

In other words, achieving this goal requires meeting several critical objectives. These include high weight percent and volume density of stored hydrogen (especially for mobile applications), favorable absorption (or adsorption) and desorption kinetics at practical temperatures and pressures, and the use of inexpensive and widely available materials. Additionally, long-term resistance to contamination by trace impurities in hydrogen streams is a crucial factor.

Table 5.4 lists some of the promising compounds for hydrogen storage.

5.3.4 Reversible Metal Hydride–Hydrogen Storage

Various metals, including titanium, iron, manganese, nickel, chromium, and their alloys, can react with hydrogen to form metal hydrides. This property is notably efficient and stable in palladium (Pd), as utilized in hydrogen sensors. Metal hydrides are also integral to nickel–metal hydride batteries found in cell phones and computers. For large-scale storage, a broader range of metals and alloys are under investigation.

Metal hydrides are particularly appealing for hydrogen storage in fuel cell-driven vehicles due to their high volumetric density. The hydrogen absorption process in metals or alloys is reversible, following the general reaction:

$$AB_2 + \chi H_2 \rightarrow AB_2H_2x + \Delta Q$$

Here, A and B represent metals (AB_2 is an alloy), and ΔQ is the heat released upon hydrogen absorption, typically characterized by the enthalpy ΔH determined

Table 5.4 Hydrogen Storage Material

Name	Formula	Percent Hydrogen	Specific Gravity	Vol. (L) to Store (1 kg H_2)	Notes
Simple Hydrides					
Liquid H_2	H_2	100	0.07	14	Cold, −252 °C
Lithium hydride	LiH	12.68	0.82	6.5	Caustic
Beryllium hydride	BeH_2	18.28	0.67	8.2	Very toxic
Diborane	B_2H_6	21.86	0.417	11	Toxic
Liquid methane	CH_4	25.13	0.415	9.6	Cold −175 °C
Ammonia	NH_4	17.76	0.817	6.7	Toxic, 100 ppm
Water	H_2O	11.19	1.0	8.9	
Sodium hydride	NaH	4.3	0.92	25.9	Caustic, but cheap
Calcium hydride	CaH_2	5.0	1.9	11	
Aluminum hydride	AlH_3	10.8	1.3	7.1	
Silane	SiH_4	12.55	0.68	12	Toxic, 0.1 ppm
Potassium hydride	KH	2.51	1.47	27.1	Caustic
Titanium hydride	TiH_2	4.40	3.9	5.8	
Complex Hydrides					
Lithium borohydride	$LiBH_4$	18.51	0.666	8.1	Mild toxicity
Aluminum borohydride	$Al(BH_4)_3$	16.91	0.545	11	Mild toxicity
Lithium aluminum	$LiAlH_4$	10.62	0.917	10 hydride	
Hydrazine	N_2H_4	12.58	1.011	7.8	Toxic, 10 ppm
Hydrogen Absorbers					
Palladium hydride	Pd_2H	0.471	10.78	20	
Titanium iron hydride	$TiFeH_2$	1.87	5.47	9.8	

Source: Fuel Cell Systems by James Laramie and Andrew Dicks

from pressure–composition–temperature isotherms. The reaction to release hydrogen provides a small amount of heat. Hydrogen absorption typically occurs in alloys within a pressurized vessel over several minutes. The vessel, containing the metal hydride, remains sealed and connects to hydrogen-consuming equipment. Refueling, akin to gasoline, involves pumping hydrogen into the car's storage material. In vehicles, the system may be slightly warmed to expedite hydrogen supply (Table 5.5).

Table 5.5 Key Properties of Metal Hydrides

Metal Hydrate Parameters for Hydrogen Storage					
Hydrate Phase	**Max. H_2 Capacity (wt%)**	**Reversible H_2 Capacity (wt%)**	**Desorption Pressure (bar)**	**T (°K)**	**Enthalpy (kJ/mol H_2)**
MgH_2	7.66	<7.0	~10^{-6}	552	74.5
VH_2	3.81	1.9	2.1	285	40.1
$MgNiH_4$	3.59	3.3	~10^{-5}	528	64.5
$TiFeH_2$	1.89	1.5	4.1	265	28.1
$ZrNiH_3$	1.96	1.1	~5×10^{-6}	573	68.6
$TiMn_{1.4}V_{0.62}H_{3.4}$	2.15	1.1	3.6	268	28.6
$ZrM_{n2}H_{3.6}$	1.77	0.9	0.001	440	53.2
$LaNi_5H_{6.5}$	1.49	1.28	1.8	285	30.8
$LaNi_{4.8}Sn_{0.2}H_{6.0}$	1.40	1.24	0.5	312	32.8

Source: United States Department of Energy, https://www.energy.gov/sites/prod/files/2015/02/f19/fcto_h2_storage_summit_motyka.pdf, Public Domain

5.3.4.1 Metal Hydride–Hydrogen Storage

The critical hydrogen-storage properties, notably total hydrogen capacity and reversibility, are typically expressed as weight percentages. However, equally important are the pressure and temperature conditions required for the release of hydrogen gas. Generally, hydrides should provide hydrogen gas within the range of 1–10 bar and over temperatures spanning from 270–600°K, depending on the heat source's availability. Research efforts are aimed at developing hydride combinations that can operate at or near atmospheric pressure and at room temperature.

Several complex metal hydrides have emerged as leading candidates, and among them, magnesium hydride (MgH_2) stands out. Particularly attractive are compounds known as alanates, which incorporate aluminum, alkali metals, and hydrogen. Examples include $LiAlH_4$ and $NaAlH_4$, which have been utilized as hydrogen carriers for many years primarily in organic hydrogenation reactions.

In addition to alanates, borohydrides with the general formula M^+BH_4, where M represents a metal, hold promise as chemical storage media. However, they currently lag in terms of energy density compared to other methods. Another avenue showing potential for reversible hydrogen storage involves metal–N–H systems.

While metal hydride systems offer the potential for greater onboard fuel capacity than compressed gas or liquid hydrogen systems, they are still in the early stages of development. Identifying a specific material with the desired hydrogen capacity, thermodynamic properties, and kinetic behavior remains an ongoing challenge in this field of research. These advanced metal hydride systems offer a promising avenue for research and development. It is crucial to realize their full potential and start transitioning away from the laboratory and into practical applications.

5.3.5 Alkali Metal Hydrides

Alkali metal hydrides offer a promising alternative to reversible metal hydrides for hydrogen production, reacting with water to release hydrogen and forming a metal hydroxide in the process. Key examples of these hydrides are listed in Table 5.4 earlier in Section 5.3.3.

Consider the system that utilizes calcium hydroxide, first documented in 1990. This process can be represented by the chemical reaction:

$$CaH_2 + 2H_2O \rightarrow Ca\left(OH\right)_2 + 2H_2$$

A commercially viable method, known as "Powerballs," employs sodium hydride reacting with water:

$$NaH + H_2O \rightarrow NaOH + H_2$$

In this system, sodium hydride is encased within polyethylene spheres approximately 3 cm in diameter. These spheres, termed Powerballs, are stored in water. They are bisected as needed, triggering the reaction between sodium hydride and water to produce hydrogen. This splitting process is meticulously controlled by a microprocessor, enabling on-demand hydrogen generation.

When compared to other methods explored thus far, this approach offers a straightforward and efficient means of hydrogen production. Sodium hydride is cost-effective and boasts a high energy density. Its potential for use in fuel cells, particularly in automotive applications, is notable due to its superior hydrogen storage capabilities relative to other methods. However, a significant drawback is the management of the corrosive hydroxide–water mixture that results from the process. Additionally, a few publications mentioned that some metallic hydrides are hydrogen deficient, so they do not hold the law of constant H_2 composition.

5.3.6 Carbon Nanostructures

Recent advancements in the field of materials science have brought carbon-based nanomaterials, particularly carbon nanotubes and fibers, to the forefront as potential solutions for efficient hydrogen storage. These materials, characterized by their diverse pore sizes, shapes, and distributions, have shown considerable promise in this application. The development in the molecular engineering of pore structures and surface chemistry in these carbon compounds has been a pivotal factor in exploring their use for hydrogen storage. One of the most notable features of these nanotubes is their ability to adsorb hydrogen at room temperature with only moderate overpressures, and then release it with minimal heating, a property that has led to significant research interest. According to the US DOE, for carbon materials to be feasible for fuel cell applications, particularly in automotive sectors, they need to store at least 6.5% of their own weight in hydrogen. Researchers at MIT created nanotube clusters capable of storing up to 4.2% of their weight in hydrogen. The NREL demonstrated that single-walled nanotubes (SWNTs) can adsorb up to 8 wt% hydrogen, especially in the presence of catalytic metal species. However, some conflicting reports indicate that room temperature storage capacities are very low.

Carbon nanotubes are typically produced in bundles and are known for their lightweight nature and high density of small, uniform, cylindrical pores. These nanotubes can be categorized into SWNTs and multiwalled nanotubes (MWNTs). SWNTs are composed of a single graphite layer, while MWNTs consist of multiple concentric graphite layers. The diameters of SWNTs vary, typically ranging from 0.671 to 3 nm, whereas MWNTs usually exhibit diameters between 30 and 50 nm. In contrast to nanotubes, carbon fibers are composed of graphite platelets stacked together, with an interlayer spacing of about 0.3355 nm.

The interaction between hydrogen and carbonaceous materials involves two primary mechanisms: van der Waals attractive forces (physisorption) and the overlap of the highest occupied molecular orbitals of carbon with the hydrogen electron, which helps overcome the activation energy barrier for hydrogen dissociation (chemisorption). The central question in this field of study is whether carbon nanotubes can store and release practical amounts of hydrogen under reasonable conditions release.

5.3.7 Other Technologies

Contemporary research in hydrogen storage is exploring several promising methods, focusing on sorbent materials capable of reversible hydrogen uptake and release. These materials are being evaluated for their effectiveness based on gravimetric and volumetric uptake capabilities. One challenge identified is that the interaction between hydrogen and carbon is too weak, whereas the metal–hydrogen bond is too strong for practical room-temperature reversible storage.

While these innovative approaches hold promise and have made significant progress, it is important to note that an ideal technology for hydrogen storage based on carbon nanostructure has not yet been realized.

5.4 Hydrogen Infrastructure

The development of a hydrogen infrastructure involves a spectrum of transportation methods to facilitate the distribution of hydrogen from production sites to end-users. These methods encompass

- Compressed pure gaseous hydrogen pipelines
- Liquid hydrogen pipelines
- Blending hydrogen with other gases or liquids, including ammonia
- Metal hydride systems
- Hydrogen extraction from blending mixtures
- Hydrogen transportation in tanks or liquid containers (Fig. 5.12)

Advancements in hydrogen production and transportation technology haven't solved a fundamental challenge: the energy used to produce hydrogen can be

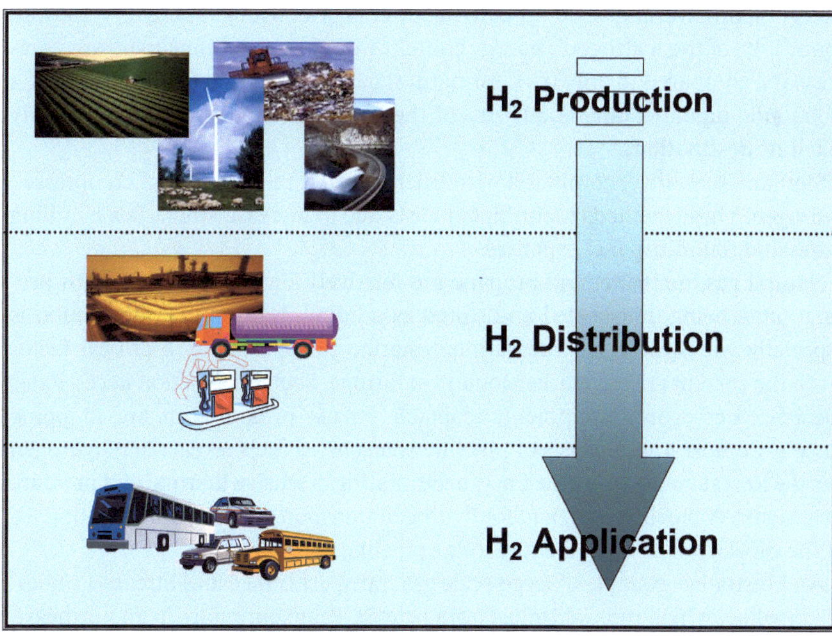

Fig. 5.12 Hydrogen Infrastructure

roughly equivalent to the energy it delivers as fuel. This creates a crucial bottleneck. Widespread adoption of hydrogen fuel cell vehicles hinges on a convenient refueling network, but such a network is unlikely to be built without a substantial number of hydrogen vehicles on the road. This creates a classic "chicken or the egg" scenario: which needs to come first?

5.5 Hydrogen Transportation via Pipeline

The United States possess an extensive network of million miles of pipelines, currently used for transporting natural gas, propane, and gasoline. This existing infrastructure presents an appealing prospect for utilizing existing pipelines to accommodate for hydrogen transportation in either compressed or liquid form.

Eliasson and Bossel have conducted a comprehensive analysis of the energy requirements for transporting hydrogen over long distances through pipelines. Their study reveals that due to hydrogen's low volumetric energy density, even at high pipeline pressures, it must be moved more rapidly than natural gas to deliver the same amount of energy. Consequently, despite hydrogen's lower viscosity, the resistance to flow significantly increases in pipelines. The analysis indicates that transporting hydrogen requires approximately 4.6 times more energy than natural gas for an equivalent energy delivery rate. Every 150 km, about 1.4% of the hydrogen's energy content would be consumed to power compressors, compared to only 0.3% for natural gas. In a hypothetical scenario of a 3,000-mile pipeline, only about 70% of the hydrogen input would successfully reach its destination.

Similar issues are encountered with truck or rail transportation of compressed hydrogen. These methods incur higher costs due to increased materials handling losses and round-trip fuel expenses.

Natural gas, methane, and propane are relatively simple to liquefy, with propane often being transported and stored as a liquid. Natural gas liquefaction is especially widespread for international marine transportation, a critical factor given the current energy market context in Europe. Such liquefaction necessitates specialized cryogenic equipment, adaptable for use in both ports and shipping vessels. Comparative analysis of pipeline-transported fuels reveals that hydrogen has the lowest volumetric efficiency, necessitating extensive thermal and pressure treatments. A pipeline designed for hydrogen transport can only carry around 8% of the capacity of an equivalent gasoline pipeline.

An illustrative example of large-scale gas transport is the Nord Stream 1 undersea pipeline, which runs 745 miles from near St. Petersburg, Russia, to northeastern Germany. This pipeline, operational from 2011 until its shutdown in 2022, had a capacity of transporting 110 billion cubic meters of natural gas annually through its four 48-in. diameter pipes. In contrast, the planned Nord Stream 2 pipeline was never realized.

Hydrogen, in comparison with other fuels, exhibits the lowest volumetric efficiency when transported by pipeline. Even under high pressure, such as 5,000 psi, a hydrogen pipeline can only carry about 12.8% of the volume achievable by an equivalent gasoline pipeline. This efficiency gap presents a significant challenge in the practical application of hydrogen as a mainstream energy carrier.

Additionally, the liquefaction of gases like natural gas, methane, and propane is relatively straightforward, with propane frequently being shipped and stored in liquid form. For international marine transportation, particularly under the current European energy market conditions, natural gas liquefaction is crucial and requires specialized cryogenic equipment at ports and on ships. Hydrogen, with its unique physical and chemical properties, presents a more complex scenario, necessitating innovative solutions for effective liquefaction and transport.

There are significant challenges associated with hydrogen transportation. High-pressure hydrogen has a propensity to leak through minuscule openings and can permeate materials that are typically impervious to natural gas. As a result, directly adapting the existing natural gas infrastructure for hydrogen delivery is not a feasible option due to the unique physical properties of hydrogen gas. Designing a land transmission system for hydrogen requires specialized considerations to mitigate energy loss, address material science challenges, and compensate for the pressure losses during transmission. Additionally, the use of high-pressure pipelines for hydrogen necessitates unique welding techniques, specialized designs for valves, pumps, sensors, materials for piping, and safety instruments to prevent the potential failures and leaks associated with high-pressure systems. To maintain operational pressure levels for natural gas delivery, compressor stations are usually placed every 50 miles along the pipeline. However, hydrogen's higher operational pressure demands more frequent compressor stations, which in turn reduces the overall energy efficiency of the system.

The issue of hydrogen-induced embrittlement in mild steel, a common material in pipeline construction, presents another challenge. This embrittlement necessitates the development of strategies to control hydrogen permeation and prevent leaks. An emerging solution is the use of fiber-reinforced polymer (FRP) pipelines for hydrogen distribution. Not only are FRP pipelines approximately 20% less costly to install compared to steel pipelines, but they also offer reduced welding expenses as FRP can be transported in longer sections.

The capability of a fuel pipeline to efficiently transport energy is directly tied to the energy density of the fuel at the operating pressure of the pipeline. The current spectrum of pure hydrogen transportation technologies encompasses three practical options: cryogenic methods, high-pressure tanks, and blending hydrogen with different media. At a pipeline pressure of 5,000 psi, hydrogen's energy content is approximately 114,285 BTU/ft^3. This value is about five times greater than at 1,000 psi and increases 10-fold at 10,000 psi.

Implementing a 5,000 psi pressure in pipelines substantially improves the energy-per-volume ratio. However, this improvement comes with the need for complex engineering solutions and significant material resources. Moreover, in densely populated areas, such adaptations may not be feasible. To accommodate the requisite high volumes of hydrogen, larger pipe diameters might be employed, but this approach presents identical engineering and urban planning challenges.

In comparison, land transportation of natural gas typically employs pipelines with diameters ranging from 30 to 48 in., operating at pressures between 500 and 5,000 psig. For instance, the previously discussed Nord Stream project exemplifies this, where gas is transported over 750 mi. underwater from Russia to Germany at pressures up to 3,191 psig, without intermediate compressor stations.

Transporting compressed hydrogen using existing gas pipelines might become a reality, but it faces considerable production and maintenance costs. These costs are due to the requirements for high pressure (up to 10,000 psi); pipe material durability; corrosion resistance; and the necessary redesign of control valves, armatures, and compressors. Creating infrastructure for liquid hydrogen transport is further complicated by prohibitive development costs and losses associated with evaporation and leakage.

Critics of hydrogen-based energy systems point out that distributing hydrogen in volumes large enough to significantly replace fossil fuels is a complex and challenging endeavor. All fuels, including hydrogen, require substantial energy for production and distribution. Current hydrogen production and transportation technologies suggest that the energy expended in producing hydrogen is comparable to that used for its compression or conversion to a cryogenic state.

Opponents also raise significant economic concerns regarding the proposed hydrogen economy. Utilizing existing gas pipelines, such as Nord Stream 2, for hydrogen transportation is conceivable but would involve considerable production and maintenance costs. These costs are linked to high pressure; leak prevention; material compatibility; and the redesign of armatures, compressors, and control instrumentation. The high initial capital costs of constructing new pipelines are a major obstacle to expanding hydrogen pipeline delivery infrastructure.

Initially, transporting gaseous hydrogen via existing pipelines seems a cost-effective option. However, numerous technical limitations necessitate innovative solutions to address pipeline transmission concerns, such as

- Losses from material penetration
- Conversion losses into liquid hydrogen

- H_2 permeation and corrosion in materials
- Cost for development
- High-pressure piping requirements
- Energy costs associated with compression
- Absence of perfect H_2 storage technology
- Pressure losses due to friction and transmission drag per mile
- High-pressure handling requirements
- Safety considerations
- Environmental considerations

The transportation of hydrogen poses unique challenges due to its specific properties and requirements for high-pressure containment. While current technologies offer viable methods such as cryogenic transport, high-pressure tanks, and blending with other media, each approach has its inherent complexities and costs. The adaptation of existing infrastructure, like natural gas pipelines, for hydrogen delivery is fraught with technical and economic challenges, including material compatibility, pressure handling, and safety concerns. Looking forward, the industry must focus on developing innovative solutions to overcome these barriers. This includes advancing material science for pipeline construction, improving compression and liquefaction technologies, and optimizing safety mechanisms. Emphasis should also be placed on cost-effective and sustainable production methods to ensure that the energy output of hydrogen as a fuel justifies the input required for its production and distribution. By addressing these issues, the hydrogen energetics field can move closer to realizing its potential as a key player in the transition to a more sustainable energy future.

5.6 Ammonia as an Energy Carrier

Ammonia's role in global carbon emissions, currently at 1.8%, is a significant concern, yet its potential as a sustainable energy carrier offers a contrasting perspective. The 2020 Royal Society report suggests that "green" ammonia, produced using low-carbon hydrogen such as blue and green hydrogen, could lead to complete decarbonization of ammonia production by 2050, aligning with global net-zero emission targets. This forward-looking vision positions ammonia as an important element in future energy strategies, especially in the realm of hydrogen energetics.

Utilizing ammonia as an energy carrier, especially for hydrogen piping and storage, is attracting attention. Its application spans from on-board hydrogen storage in fuel cell vehicles to off-board storage solutions, underlining its versatility. Ammonia boasts a significant hydrogen storage capacity of 17.6 wt%, compared to approximately 5% for pressurized gaseous hydrogen.

The conversion of ammonia to hydrogen, particularly through the sodium amide process, is viable for both combustion and as a fuel source in proton exchange membrane fuel cells.

The challenge in harnessing ammonia and hydrogen as alternative energy carriers lies not only in their physical properties but also in the necessary technological advancements and global commitment to sustainable practices. The future of energy, with a focus on reducing emissions and increasing efficiency, will significantly depend on the successful integration of these alternative carriers into the existing energy infrastructure. As the world moves toward greener energy solutions, the role of ammonia and hydrogen in achieving these goals becomes increasingly pivotal.

Directly employing ammonia as an onboard fuel processor entails various technical challenges. These include the considerable energy requirement to release hydrogen from ammonia, the need for substantial equipment in terms of mass and volume, and concerns regarding user health, particularly due to the risk of PEMFC poisoning from trace amounts of ammonia (>0.1 ppm).

These include maintaining high operating temperatures (over 500 °C), ensuring the longevity and reliability of catalysts and other components in the presence of impurities, managing prolonged start-up times, and meeting stringent purification standards to prevent fuel cell poisoning. The complexity of the system, coupled with concerns about energy efficiency and cost, poses significant hurdles. Additionally, the reactor's considerable weight and volume, typically ranging from 2,000 to 5,000 kg and occupying about 3,000–6,000 L, present practical challenges in implementation.

In the context of hydrogen storage and transportation, liquid organic hydrogen carriers (LOHCs), particularly ammonia, offer an alternative to traditional methods. Ammonia enables efficient hydrogen absorption and release through chemical reactions. This positions ammonia as a leading LOHC option, promising to revolutionize hydrogen energetics.

The energy density of liquid ammonia is approximately 11.5 MJ/L, about a third of diesel's density. This offers the opportunity to convert ammonia back into hydrogen for powering hydrogen fuel cells, or for direct use in high-temperature solid oxide direct ammonia fuel cells, providing efficient power sources free from greenhouse gas emissions.

Ammonia, when compared to hydrogen, is more energy-efficient and can be produced, stored, and delivered at a considerably lower cost. This is particularly advantageous over hydrogen, which requires compression or cryogenic liquid storage. The emergence of blue ammonia in the energy market adds another dimension, offering possibilities for its conversion back into blue hydrogen or direct use as a fuel in gas turbines, facilitating cleaner power generation.

In the broader spectrum of hydrogen compounds, binary hydrogen compounds, including hydrides, play a vital role. These compounds can be categorized into several types, each with unique properties and applications in the energy sector (Table 5.6).

Table 5.6 Hydrogen Storage Options vs Cost

Storage options	Cost comparison
Liquid hydrogen	Very expensive
Compressed hydrogen	Expensive
Liquid organic hydrogen carriers (LOHCs)	Expensive, not fully developed
Metal hydrates	Low efficiency, complicate
H_2 tanks transportation	Available technology

When evaluating energy content among commonly used fuels, notable comparisons emerge. One gallon of gasoline, for instance, is equivalent to 380 SCF of hydrogen (10 Nm^3), 113 SCF (3 Nm^3) of natural gas, or 8.3 lb. of coal. These comparisons highlight the varying energy densities and underscore the challenges in efficiently storing and transporting hydrogen.

5.6.1 Compressed Hydrogen

Hydrogen, known for its notably low volumetric energy density, necessitates compression for effective pipeline transportation. However, even when compressed to 800 bar, hydrogen's efficacy as an energy carrier remains limited. Eliasson and Bossel's analysis reveals that the energy required for compressing hydrogen from 1 to 200 bar consumes about 7.2% of its heating value. They further estimate that compression to 800 bars could consume up to 10% of the energy that hydrogen is capable of producing. This significant energy requirement for compression is substantial compared to other compressed gases, posing a major challenge to hydrogen's current economic viability as an energy carrier. This high energy cost for compression impacts hydrogen's role in the energy sector, particularly when evaluating its efficiency and cost-effectiveness in comparison to alternative energy sources. As a result, the development of more energy-efficient compression methods and the exploration of other forms of hydrogen storage and transport remain crucial areas for research and innovation in the field of hydrogen energetics.

5.6.2 Liquid Hydrogen

In the realm of hydrogen energetics, transporting hydrogen over short distances often involves its liquid form, utilizing vehicles such as insulated rail cars or trucks fitted with industrial-sized Dewar or vacuum containers. These specialized vehicles operate much like large-scale vacuum flasks, occasionally enhanced with cooling systems to keep the hydrogen at cryogenic temperatures.

This method of transportation is akin to how natural gas is liquefied for international marine transport, requiring specific cryogenic equipment for both shipping and receiving ports. However, liquid hydrogen presents its own set of challenges, most notably its very low density. At 20°K, it has a density of only 4.43 lb./ft^3, which is about 7% of the density of water. This low density makes liquid hydrogen less favorable compared to other liquefied fuels in terms of volumetric efficiency. For perspective, a cubic foot of hydrogen contains 274,900 BTU of combustion energy, which is significantly lower than the 922,500 BTU in a cubic foot of gasoline (about 45 lb.) and the 784,000 BTU in propane (around 36 lb.).

The energy requirements for hydrogen liquefaction are considerable. Eliasson and Bossel's analysis sheds light on this aspect, pointing out that the cooling process for hydrogen is extraordinarily energy-intensive. With a Carnot efficiency of only 7–12%, it is estimated that even in large-scale, efficient plants, about 40 MM kJ/kgH_2, or roughly 30% of the gas's heating value, is consumed in the liquefaction process. Moreover, the extremely low temperatures required for transporting hydrogen in liquid form necessitate both pressurization and refrigeration to prevent boil-off. This is compounded by the challenge of ambient heating, which demands substantial re-refrigeration along extended pipelines. In some cases, the energy expended to transport hydrogen can exceed the energy it delivers. These challenges are compounded when considering pipeline installation in urban areas, where space and safety are primary concerns.

The notion that hydrogen, being a gas at atmospheric pressure, can be easily transported from its production site to the point of use is overly simplistic. In reality, the compression of hydrogen to volumes sufficient for practical applications is an energy-intensive process. Public apprehension toward high-pressure vessels further complicates this issue. Likewise, the liquefaction of hydrogen prior to transportation and distribution presents its own set of problems, including the costs associated with processing, logistics, and public acceptance.

While liquid hydrogen presents a viable option for short-distance transportation, the energy-intensive nature of its liquefaction, combined with logistical and public acceptance challenges, underscores the need for continued innovation in hydrogen transport technologies. Future advancements in this field must address both the efficiency of transport methods and the broader acceptance of these technologies to fully realize hydrogen's potential as a sustainable energy carrier.

Analyzing the physical properties of hydrogen in comparison to other fuels such as methane (CH_4) reveals significant challenges in its storage and transportation. Hydrogen's volumetric energy density is markedly lower, necessitating a much larger volume to store an equivalent amount of energy as that of conventional fuels like gasoline or natural gas.

The broader implications of hydrogen's physical properties are significant for its role as an energy carrier. Its massive volume relative to energy output impacts the feasibility of its use in existing infrastructure and raises questions about its economic viability in comparison to traditional fuels. The need for innovation in hydrogen storage and transportation is evident, with a focus on developing more efficient, cost-effective methods that can integrate seamlessly into existing systems.

While hydrogen holds promise as a clean energy source, its current limitations in storage and transportation efficiency necessitate continued research and development. Advancements in this field are crucial to overcome the challenges posed by hydrogen's physical properties and to unlock its potential as a sustainable and economically viable alternative to fossil fuels.

5.7 Blending Hydrogen in Natural Gas Pipelines

While traditional hydrogen production and delivery methods are relatively well-established, hydrogen blending in natural gas pipelines depends significantly on the specific characteristics of the pipeline system. The Gas Technology Institute (GTI) assessment detailed the implications of hydrogen blending, considering the distinct features of the US natural gas pipeline system. This approach underscores the need to tailor hydrogen storage and delivery methods to the unique requirements of existing infrastructure, ensuring efficient and sustainable integration of hydrogen into the energy sector (Table 5.7).

5.7.1 Methanol Transportation as a Comparison

The other potential future fuel is methyl alcohol (methanol). Easily produced from natural gas, methanol burns cleanly without soot and has a high-octane number. However, it is toxic, similar to ethyl alcohol, and aggressively corrodes metals, necessitating chemically stable containers for storage and transportation. Its interaction with water also presents challenges. However, from an environmental standpoint, a methanol spill may be less impactful, potentially going unnoticed by environmentalists. The fuel comparison is done by using Specific Heat Energy Value – the amount of heat generated by burning unit mass of fuel to increase its temperature by one degree (see Table 3.2 in Chapter 3).

Table 5.7 Higher Heating Value (HHV) per Liter for Various Fuels

Fuel	Pressure (bar/psi)	HHV (MJ/L)
Gasoline (liquid)	Ambient	33.5
Methane (gas)	800/11,600	32.5
Propane (liquid)	Ambient	25.2
Methanol (liquid)	Ambient	17.5
Hydrogen (gas)	800/11,600	10.2
Hydrogen (liquid)	~200 psi at −240 °C	10.0

5.7.2 Petroleum Transportation as a Comparison

To contextualize, consider petroleum transportation in the United States. In 2000, three-fourths of crude petroleum and nearly two-thirds of petroleum product ton-miles were transported by pipeline, with more than 200,000 miles of oil pipeline operating across all 50 states. Motor carriers, which transport about 31% of petroleum products (mainly gasoline, diesel fuel, and jet fuel), offer a contrast to pipeline transport.

Utilizing such high pressures in pipelines necessitates larger diameters, leading to increased engineering and equipment costs, especially when considering the installation in urban areas.

The integration of hydrogen as a primary energy carrier is a subject of growing interest and debate, particularly in the context of replacing fossil fuels. However, distributing hydrogen in the volumes needed for widespread adoption presents significant challenges. Producing an amount of hydrogen equivalent in energy to a gallon of gasoline requires substantial energy, raising questions about the economic feasibility of using existing pipeline infrastructures for hydrogen transportation.

Current technologies for hydrogen production and transportation suggest a concerning trend: the energy required to produce hydrogen may exceed the energy it can offer as a fuel. This poses a dilemma for hydrogen's role as a sustainable energy source.

Hydrogen's small atomic size allows it to seep through the crystal lattice of most metals and nonmetals, leading to gradual leakage into the atmosphere. This property necessitates specific safety measures, particularly in enclosed spaces like garages. For vehicles powered by hydrogen, forced ventilation systems are essential to prevent the accumulation of gas and potential self-ignition once a certain concentration is reached.

While hydrogen holds promise as a clean and sustainable energy source, its transportation via pipelines poses significant engineering and safety challenges. Addressing these issues requires innovative solutions in pipeline design, material science, and safety protocols to harness hydrogen's full potential as an energy carrier.

5.7.3 Realistic Approaches for Hydrogen Transportation

Given these constraints, realistic approaches to hydrogen transportation are being explored:

- *Localized Production, Storage, and Dispersal*: This approach focuses on producing, storing, and distributing hydrogen at a single location. It offers a streamlined, efficient method that reduces the complexities associated with long-distance transportation.

- *LOHCs*: Substances like ammonia are used as LOHCs, providing an alternative for hydrogen storage and transport, potentially overcoming some of the challenges associated with gaseous hydrogen.
- *Hydrogen Tanks for Distribution*: These tanks are utilized for distributing hydrogen and for individual usage in mobility devices, offering a more direct and localized approach to hydrogen delivery.
- *Hydrogen Piping*: The transportation of hydrogen through pipelines, while challenging, remains a potential method for distribution, albeit with considerable infrastructure requirements.

To better understand hydrogen's challenges, it is informative to compare its energy content with that of other fuels. As mentioned earlier, one gallon of gasoline is equivalent to 380 SCF of hydrogen (10 Nm^3), 113 SCF (3 Nm^3) of natural gas, or 8.3 lb. of coal. This comparison highlights that hydrogen's volume, for a given amount of energy, is significantly larger than that of gasoline, natural gas, or any other conventional fuel. This disparity has profound implications for storage and transportation.

The transition to hydrogen as a key energy source is fraught with challenges, primarily due to its energy-intensive production and the logistical complexities of storage and distribution. While current technologies present obstacles, innovative approaches are being explored to make hydrogen a viable, sustainable energy source. The successful integration of hydrogen into the energy mix will require overcoming these technical hurdles, underscoring the need for continued research and development in this area.

The effective integration of hydrogen as an energy source is pivotal in the shift toward cleaner energy solutions. The chemical and petroleum industries provide a practical model for hydrogen utilization, primarily using it close to their production sites. This approach, which combines production and usage in a single location, offers an efficient and cost-effective solution to the challenges of hydrogen transportation.

Adopting this model for hydrogen means constructing hydrogen-fueling stations near production facilities. This setup can integrate small-scale hydropower and nuclear energy sources, alongside renewable options like solar and wind energy. This approach not only reduces the need for extensive and expensive hydrogen piping across the country but also maximizes the use of local energy resources.

Localizing hydrogen production, storage, distribution, and consumption emerges as a viable strategy to mitigate the high costs associated with hydrogen piping. This approach not only addresses the economic challenges but also leverages the potential of hydrogen as a clean and sustainable energy source. By optimizing the hydrogen supply chain, we can enhance its accessibility and usability, paving the way for its increased adoption in various energy applications (Table 5.8).

Table 5.8 Hydrogen Storage Options vs. Cost

Storage options	Cost comparison
Liquid hydrogen	Very expensive
Compressed hydrogen	Acceptable option
Liquid organic hydrogen carriers (LOHCs) as ammonia	Expensive, not developed yet
Metal hydrates	Low efficiency, complicate
H_2 tanks transportation is 350 and 700 bar for distribution and individual usage for mobility devices via road lines	Available technology

5.8 Hydrogen Bonding

Hydrogen bonding, an essential concept in chemistry, refers to the electrostatic attraction between a hydrogen atom in one polar molecule (like water) and a small electronegative atom (such as oxygen, nitrogen, or fluorine) in another molecule. This type of bonding forms due to the dipole–dipole interaction, where the hydrogen atom, covalently bonded to a highly electronegative atom, attracts another electronegative atom nearby. For example, in water molecules (H_2O), hydrogen atoms are covalently bonded to the more electronegative oxygen atoms, creating a network of hydrogen bonds.

Beyond its common forms as a compressed gas or liquid, hydrogen can be stored in alternative ways, such as blending with natural gas or diesel. These methods are increasingly being considered for their potential in effective hydrogen transportation and storage.

The blending of hydrogen with other gases is being researched as a way to improve infrastructure and reduce greenhouse gas emissions. Adapting the existing natural gas delivery infrastructure to accommodate hydrogen is one approach. Modifying pipelines to carry a blend of natural gas and hydrogen (up to about 15% hydrogen) could require relatively minor modifications. However, adapting these pipelines for pure hydrogen delivery necessitates more extensive changes due to hydrogen's distinct physical and chemical properties.

When comparing the energy content of fuels, hydrogen's combustion heat is three times greater than gasoline by mass, due to its lightweight nature. However, when measured by volume, hydrogen's energy yield is lower – a cubic meter of hydrogen provides 10,800–12,770 kJ, while a cubic meter of methane delivers 35,840–39,860 kJ. Hexane vapor, representing gasoline, yields over 160,000 kJ/m^3. This significant difference highlights hydrogen's suitability for fuel cells, where it can be directly converted into electricity using catalysts, such as palladium, rather than being burned.

The combustion of hydrogen occurs at very high temperatures, necessitating materials in combustion chambers that can withstand up to 3,000 °C. Drawing from personal involvement in hydrogen utilization for internal combustion processes in the late 1980s, it was found that the primary issues were managing leakage and handling the high temperatures required for hydrogen combustion. These challenges point to the need for advanced materials science and innovative engineering solutions to effectively utilize hydrogen in various energy applications.

Hydrogen's unique properties, from its bonding behavior to its storage and energy efficiency, present both opportunities and challenges in its role as an energy carrier. The future of hydrogen technology lies in overcoming these challenges through continued research, development, and innovation in materials science, engineering, and infrastructure adaptation.

Derived from abundant resources like water and electricity, hydrogen is poised to disrupt the current energy monopoly. Countries such as Israel and Germany are at the forefront of developing innovative hydrogen gel technologies. This gel facilitates easy storage, transportation, and quick rechargeability of hydrogen fuel cells. A notable advancement in this field is the creation of a new silicon-based powder, enhancing the safety and efficiency of hydrogen fuel transportation.

Comparing the costs of gasoline and hydrogen directly is not straightforward due to differences in engine efficiencies. Traditional gasoline engines, especially in urban settings with frequent stops, have an efficiency of about 18–23%. In contrast, Toyota's latest hydrogen-powered vehicle achieves an impressive 83% efficiency with its hydrogen fuel cell, allowing it to travel over 200 km/kg of hydrogen.

Electric cars, often charged in the evenings, can strain power networks. In contrast, hydrogen-powered vehicles, particularly those using fuel cells, essentially operate as self-charging power plants. Coupled with the emerging hydrogen gel technology, this represents a comprehensive and sustainable energy solution.

5.9 Hydrogen Extraction from Blended Mixtures

Extracting hydrogen from blended mixtures is a critical process in the field of hydrogen energetics, involving sophisticated technologies that balance efficiency with energy consumption and spatial requirements.

5.9.1 Pressure Swing Adsorption (PSA) Technology

PSA is a prominent method for hydrogen recovery, particularly effective for low hydrogen concentration mixtures. The NREL notes that for a 10% hydrogen concentration with an 80% recovery factor, the estimated extraction cost ranges from \$0.3 to \$1.3/kg. This variation is influenced by economies of scale, with system sizes ranging from 1,000 to 100 kg/day. PSA systems designed for low hydrogen concentrations, such as 20% mixtures, must be proportionately larger to accommodate the volume of impurities. Economically, PSA is viable mainly at pipeline pressure reduction stations, where existing pressure gradients assist in hydrogen separation. In scenarios lacking such pressure drops, the energy and capital investment for compressing and reinjecting depleted hydrogen gas back into pipelines becomes substantial.

5.9.2 Membrane Separation Technology

Membrane separation is effective with higher hydrogen concentrations and is capable of achieving very high purity levels, thanks to the selectivity of certain membrane technologies. Typically employed in industrial settings for bulk hydrogen recovery, these applications often do not require extremely high purity. Lower concentration operations may necessitate a higher pressure differential across the membrane, meaning substantial volumes of nonhydrogen gases must be compressed at high pressures for effective hydrogen separation. This technology is particularly apt for high-pressure transmission pipelines, where existing pipeline pressures allow for significant hydrogen recovery.

5.9.3 Electrochemical Hydrogen Separation (EHS)

EHS, or hydrogen pumping, offers a more intricate approach for bulk hydrogen recovery. Currently, two main technologies are used: Nafion-based membrane systems and polybenzimidazole (PBI) systems. Nafion-based systems have a longer development history and are more technologically mature, but PBI systems are increasingly preferred due to their lower compression requirements. In a 1,000-psi pipeline, both the product gas and hydrogen are output at approximately 1,000 psi, minus specific process pressure drops. While EHS reduces the need for pressurization, it introduces increased complexity and has not yet reached the maturity level of PSA or certain membrane technologies.

In the realm of hydrogen energetics, electrochemical hydrogen pumping necessitates the use of water for its operation, incorporating a humidification system. Conversely, gases in pipelines must be dry, requiring a water-removal system downstream. This juxtaposition presents unique challenges in hydrogen extraction and processing.

5.9.4 Cost Analysis of Hydrogen Extraction

For a mixture with 10% hydrogen concentration and an 80% recovery factor, PSA extraction from a 300 psi pipeline costs between \$3.3 and \$8.3/kg of hydrogen, varying with recovery rates from 1,000 to 100 kg/day. For a 20% concentration, the extraction cost ranges from $2.0 to $7.4/kg for the same recovery rates. These costs are highly relative to the competitive hydrogen cost goal of $2 to $4/kg, targeted for Fuel Cell Electric Vehicle (FCEV) markets. However, increasing the hydrogen concentration to 20% can reduce costs per kilogram by approximately 10%. Thus, PSA extraction could become a more economically viable component of the total delivered hydrogen cost, particularly if extraction occurs at a pressure-reduction facility.

6

Hydrogen Production: Current Practices and Emerging Technologies

Chapter 6 examines hydrogen production. As of 2020, about 95% of hydrogen comes from fossil fuels, predominantly through natural gas steam reforming, partial oxidation of hydrocarbons, and coal gasification. These methods rely mainly on natural gas, oil, and coal, highlighting the dependency on fossil fuels for hydrogen production.

The industry, however, is shifting toward sustainable practices. Technologies like biomass gasification and zero-CO_2-emission methane pyrolysis are emerging, alongside water electrolysis, which can utilize renewable electricity sources like solar power. Additionally, carbon capture and storage (CCS) advancements are improving the environmental profile of traditional methods by efficiently separating carbon dioxide from natural gas and other hydrocarbons.

This chapter provides an overview of these hydrogen production methods, their environmental impacts, and their roles in transitioning to sustainable energy solutions, offering insights into the current state and future directions of hydrogen production.

6.1 Hydrogen Production from Fossil Sources

6.1.1 Steam Methane Reforming (SMR)

SMR stands as a primary method for hydrogen production from natural gas, predominantly methane (CH_4). This technique holds the status of being the most economical source of industrial hydrogen, accounting for nearly half of the global hydrogen supply. The SMR process involves heating natural gas to a temperature range of 700–1100 °C (1,292–2,012 °F), in the presence of steam and a nickel catalyst. This endothermic reaction leads to the decomposition of methane, resulting in the formation of carbon monoxide and molecular hydrogen (H_2).

Hydrogen Energetics: A Future Clean Energy Carrier, First Edition. Roman J. Press.

6.1.1.1 Chemical Equation and Purpose

The primary objective of SMR is hydrogen production. The process can be summarized by the following chemical equilibrium:

$$4CH_4 + O_2 + 2H_2O \rightarrow 10H_2 + 4CO$$

Following this, the carbon monoxide produced is further processed with steam over iron oxide (or other oxides) to undergo a water-gas shift reaction, producing additional hydrogen. However, a significant drawback of this method is the emission of CO_2, CO, and other greenhouse gases. The environmental impact varies with the feedstock quality, with each ton of hydrogen potentially resulting in 9–12 tons of CO_2 emissions. This aspect of SMR underscores the importance of CO_2 capture technologies (Fig. 6.1).

6.1.1.2 Process Description

The SMR process begins with the high-temperature steam reacting with methane in an endothermic reaction to yield syngas:

$$CH_4 + H_2O \rightarrow CO + 3H_2$$

Subsequently, at a lower temperature (approximately 360 °C or 680 °F), an exothermic water–gas shift reaction occurs:

$$CO + H_2O \rightarrow CO_2 + H_2$$

$$CH_4 \rightarrow C + 2H_2$$

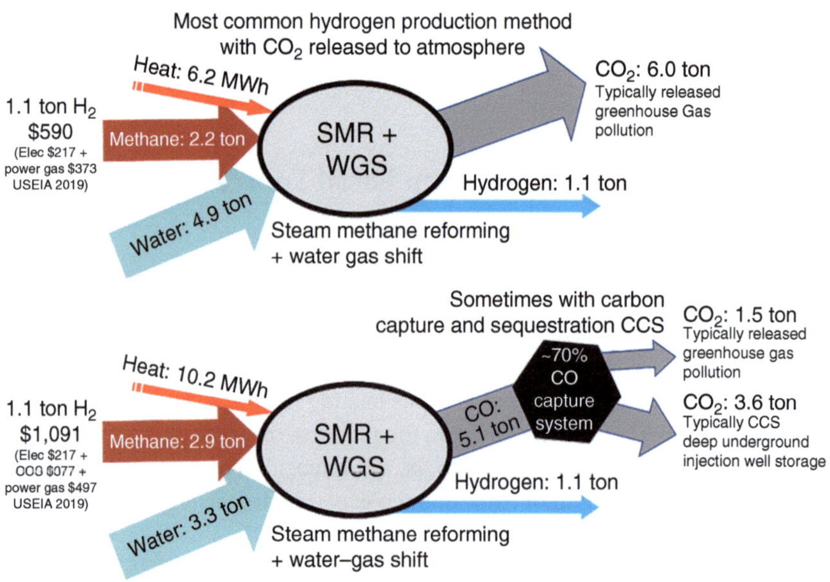

Fig. 6.1 Steam Methane Reforming. *Source:* Parent55/Wikimedia Commons/CC BY 1.0

This stage involves the oxidation of CO to CO_2 by stripping an oxygen (O) atom from additional water (steam). This oxidation not only aids in maintaining the reaction but also provides the necessary energy. Additional heat is typically supplied by burning a portion of the methane.

6.1.2 Methane Pyrolysis

6.1.2.1 Process Overview

Methane pyrolysis, another hydrogen production method from natural gas, offers a more environmentally friendly alternative. This process involves a single-step hydrogen separation using a molten metal catalyst in a bubble column. Recognized for its potential in low-cost hydrogen production, methane pyrolysis is being evaluated for scalability and operational efficiency at large scales (Fig. 6.2).

6.1.2.2 Temperature and Variations

The procedure is executed at higher temperatures, typically around 1065 °C (1950 °F). Alternative forms of methane pyrolysis, such as thermo-catalytic decomposition, can operate at lower temperatures (600–1000 °C), depending on the catalyst used.

6.1.2.3 Chemical Reaction and Environmental Impact

The chemical reaction can be represented as

$$CH_4(g) \rightarrow C\,(s) + 2H_2(g)\ \Delta H^\circ = 74.8\ \text{kJ/mol}$$

A notable advantage of this method is the production of industrial-grade solid carbon as a by-product, which can be utilized as a manufacturing feedstock or safely landfilled. This approach ensures that no atmospheric pollution occurs and mitigates the risk of groundwater contamination.

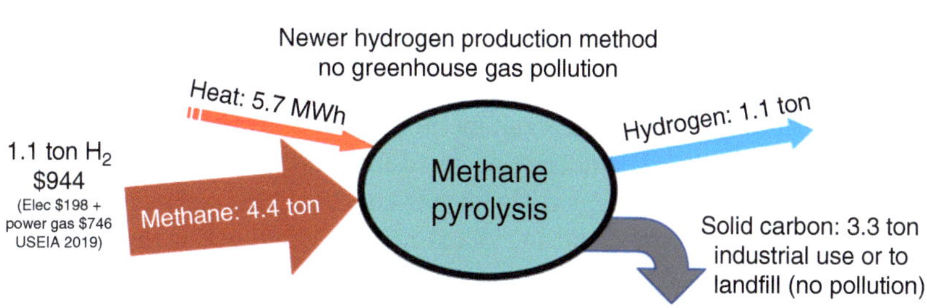

Fig. 6.2 Methane Pyrolysis. *Source:* Parent55/Wikimedia Commons/CC BY 1.0

6.1.3 Coal Gasification

Coal gasification is a prevalent method for hydrogen production, involving the heating of coke, natural gas, methane, or other light hydrocarbons (such as ethane or propane) with steam in the presence of a catalyst. The process can be represented by the following reactions:

- For coke: $C\ (coke) + H_2O\ (1100–1300\,°C) \rightarrow CO + H_2$
- For methane: $CH_4 + H_2O\ (700–925\,°C) \rightarrow CO + 3H_2$

This endothermic process requires heat, typically supplied by burning fossil fuels.

6.1.3.1 Water–Gas Shift Reaction

In both scenarios, additional hydrogen is produced via the water–gas shift reaction. This reaction involves passing carbon monoxide (CO) and steam over iron oxide or cobalt oxide at high temperatures (around 400 °C), resulting in the formation of carbon dioxide (CO_2) and hydrogen (H_2): $CO + H_2O \rightarrow CO_2 + H_2$.

6.1.3.2 Hydrogen Purification

Subsequently, a mixture of H_2 and CO_2 is directed to a gas purifier. Here, hydrogen is separated from CO_2 through various methods, including pressure swing absorption, wet scrubbing, or membrane separation.

6.1.3.3 Commercial Coal Gasification Processes

Two main commercial coal gasification processes are in use: one operates at atmospheric pressure, and the other at approximately 55 bar. Inorganic materials left after coal gasification are removed as molten slag at the reactor's bottom.

6.1.3.4 Comparison with Steam Reforming of Natural Gas

Contrasting this with the steam reforming of natural gas, another method to produce hydrogen, we observe differences in efficiency and environmental impact. Currently, steam reforming remains the most cost-effective method, accounting for about half of the world's hydrogen production (see Fig. 6.3). However, it poses environmental challenges. CO and CO_2 are major by-products, and sulfur released from raw materials forms sulfur and nitrogen compounds. These emissions necessitate environmentally friendly treatment methods.

Furthermore, the global dependency on fossil fuels for hydrogen production, including natural gas and coal, is a significant concern. This reliance is often viewed as a major source of regional and global conflict, emphasizing the need for sustainable energy development.

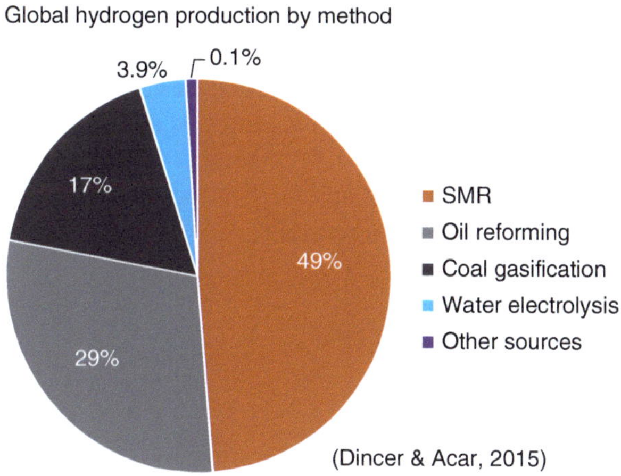

Fig. 6.3 Industrial Hydrogen Production Methods. *Source:* Alexpreeves/Wikimedia Commons/CC BY-SA 4.0

6.1.4 Partial Oxidation of Hydrocarbons

6.1.4.1 Fundamentals of the Process

Partial oxidation of hydrocarbons is a process that transforms heavy hydrocarbons (referred to as "feedstock"), such as natural gas, naphtha, petroleum coke, or coal, into a syngas mixture containing hydrogen (H_2), carbon monoxide (CO), and carbon dioxide (CO_2). This transformation is achieved using superheated steam and a reduced amount of oxygen. External energy, often obtained through the combustion of the feedstock itself, is necessary for this process.

6.1.4.2 Enhancing Hydrogen Content

To augment the hydrogen content in the produced syngas, the mixture of H_2, CO, and CO_2 undergoes the water–gas shift reaction. This reaction further converts CO and water into additional hydrogen and CO_2. The overall efficiency of partial oxidation processes is approximately 50%.

6.1.4.3 Application in Heavy Hydrocarbons

Heavy hydrocarbons, which are unsuitable for catalytic steam reforming, can effectively undergo hydrogen production through partial oxidation. This involves the partial combustion of a fuel–air or fuel–oxygen mixture, resulting in syngas rich in hydrogen and carbon monoxide. The water–gas shift reaction subsequently plays a crucial role in increasing hydrogen and carbon dioxide yields.

The partial oxidation reaction involves the partial combustion of a substoichiometric fuel–air or fuel–oxygen mixture in a reformer or a partial oxidation reactor. Two primary variants exist: thermal partial oxidation (TPOX) and catalytic partial oxidation (CPOX). The generalized chemical reaction is expressed as $C_nH_m + {}^n/_2\ O_2 \rightarrow n\ CO + {}^m/_2\ H_2$.

For illustration, consider the idealized reactions for heating oil and coal, assuming compositions of $C_{12}H_{24}$ and $C_{24}H_{12}$, respectively:

- For heating oil: $C_{12}H_{24} + 6\ O_2 \rightarrow 12\ CO + 12\ H_2$
- For coal: $C_{24}H_{12} + 12\ O_2 \rightarrow 24\ CO + 6\ H_2$

6.1.5 Petroleum-Refining Operations

6.1.5.1 Hydrogen Recovery in Refining Processes

In the realm of petroleum-refining operations, hydrogen emerges as a vital by-product, recovered from various refinery and chemical streams. These streams typically include purge gas, tail gas, and fuel gas. One illustrative example of hydrogen recovery occurs during the conversion of hexane, a hydrocarbon with a low octane rating, into benzene, a hydrocarbon boasting a higher octane rating. The chemical reaction underlying this transformation is $C_6H_{14} \rightarrow C_6H_6 + 4\ H_2$.

This reaction not only alters the hydrocarbon structure but also releases hydrogen, demonstrating an integrated approach to refining and hydrogen production.

6.1.5.2 Environmental Impact of Conventional Hydrogen Production

While hydrogen is a crucial component in various industrial processes, the conventional methodologies for hydrogen production, particularly those relying on fossil raw materials, are not without significant health and environmental repercussions. These impacts span a wide range, including air and water pollution, contributions to ozone layer depletion, and exacerbation of global warming. The reliance on fossil fuels in traditional hydrogen production methods underlines the urgency for environmentally conscious practices in industrial operations.

6.1.5.3 Ongoing Research and Development Efforts

In response to these environmental challenges, the scientific community is actively engaged in researching and developing new methods for hydrogen production. These novel methods are focused on minimizing, if not entirely eliminating, negative environmental impacts. The goal of this ongoing research is to pioneer techniques that are not only efficient and scalable but also harmonious with environmental sustainability. By innovating cleaner production methods, the industry aims to reconcile the increasing demand for hydrogen with the imperative to protect and preserve our planet's ecosystem.

6.2 Hydrogen Production from Renewable Sources

6.2.1 Green Hydrogen Production

Green hydrogen (GH_2), characterized by its production through zero-emission renewable energy sources, represents a significant leap toward sustainable energy. This form of hydrogen is generated via the electrolysis of water – splitting water molecules into hydrogen and oxygen using electricity derived from clean sources such as wind, solar, nuclear, hydro, wave, or tidal power. In this process, two electrodes connected to a power source are immersed in water, and a direct current is applied. Currently, green hydrogen contributes to less than 0.1% of total hydrogen production.

6.2.1.1 Electrolysis and Its Applications

A particularly effective method for producing hydrogen gas is through dedicated electrolyzers. The hydrogen generated can be directly used in fuel cell systems to produce electricity. More commonly, however, hydrogen gas is stored and transported to distant locations for use in fuel cells. Hydrogen and oxygen produced from electrolysis can be stored separately in pressure tanks or within other mediums like metal hydrides, making hydrogen a versatile energy carrier that can be transported and utilized according to specific needs and schedules.

The economic advantages of storing energy as hydrogen include emissions reduction, improved utilization efficiency, enhanced operational conditions, and the relative simplicity of energy transport to required sites.

In contrast, hydrogen production from fossil fuels, known as gray hydrogen, invariably results in greenhouse gas emissions. If substantial carbon dioxide emissions are captured during the process, the resultant hydrogen is termed blue hydrogen, while hydrogen derived from coal is referred to as brown hydrogen.

6.2.1.2 Challenges and Potential of Green Hydrogen

The main barrier to the widespread adoption of green hydrogen is its high production cost. A price point of $2/kg is seen as a potential tipping point for green hydrogen to compete against other fuel sources. Notably, it offers a more cost-effective solution for long-term renewable energy storage compared to alternatives like pumped-storage hydroelectricity or batteries. The anticipated decrease in renewable energy costs with global decarbonization efforts is expected to make green hydrogen increasingly viable.

Green hydrogen, when blended with natural gas, can be utilized to produce green ammonia, a key component in fertilizer manufacturing. In the foreseeable future, green ammonia is projected to become cost-competitive with conventionally produced gray ammonia.

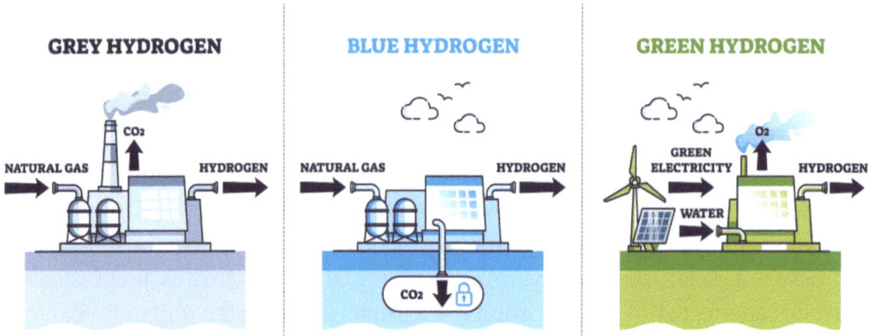

Fig. 6.4 Gray, Blue, Green Hydrogen Production Structure. *Source:* Adobe Stock

6.2.1.3 Environmental Impact

As highlighted by the International Energy Agency (IEA), adopting green hydrogen production could prevent the annual emission of approximately 830 million tons of CO_2 currently produced from fossil fuels. To replace all gray hydrogen globally, an estimated 3,000 TWh/year from new renewables would be required – equivalent to the current energy demand of Europe (Fig. 6.4).

This excess, often available at a reduced cost, presents an ideal opportunity for hydrogen production.

6.2.2 Blue Hydrogen Production

Blue hydrogen is mainly produced from natural gas by using steam reforming, which brings together natural gas and heated water in the form of steam. The output is hydrogen, but carbon dioxide is also produced as a by-product. Blue hydrogen definition also included the terms "low-carbon hydrogen," as the steam reforming process does not actually avoid the creation of greenhouse gases. The carbon dioxide that is produced during this process is then captured and stored, while the hydrogen is used for various applications.

6.2.3 Pink Hydrogen

The production of hydrogen via nuclear energy, also known as pink hydrogen, is often categorized as a subset of green hydrogen. Nuclear power is particularly notable for its independence from oil-producing sources, its minimal generation of greenhouse gases, and its political independence from offer sources. One unique advantage of nuclear plants is their ability to generate excess electricity during off-peak hours when the demand from grid consumers is low.

There are three primary methods employed to produce hydrogen using nuclear energy: traditional electrolysis, high-temperature steam electrolysis, and thermochemical water splitting cycles. Light-water reactors (LWRs), a type of thermos-nuclear reactor, unlike coal or SMR processes, produce hydrogen that does not require cleaning for chemical impurities. In the United States, 103 LWRs are located across 64 sites in 31 states, contributing to approximately 20% of the nation's electricity needs. Utilizing this off-peak electricity output for hydrogen production not only optimizes energy use but also bolsters domestic hydrogen production capabilities.

Furthermore, there are additional opportunities to integrate electricity generation with hydrogen production and ocean water desalination. This multifaceted approach can enhance the efficiency and sustainability of these processes.

Pink hydrogen, derived from nuclear energy, mirrors the production methods and advantages outlined in blue hydrogen production. Its production during off-peak hours utilizing excess electricity remains a key strategy. Moreover, there is an intriguing prospect of directly applying nuclear heat to split hydrogen from water.

High-temperature (950–1,000 °C) gas-cooled nuclear reactors, such as high-temperature gas-cooled reactors (HTGRs), are being explored for their potential in thermochemically splitting hydrogen from water. Ongoing research in HTGR technology aims to eventually make hydrogen supply from this method cost-competitive with natural gas steam reforming. As of now, the Swedish Oskarshamn Nuclear Power Plant is a notable example, producing pink hydrogen for commercial use.

6.2.3.1 The Future of Nuclear-Generated Hydrogen

This approach to hydrogen production not only diversifies the energy portfolio but also aligns with global decarbonization efforts. The dual use of nuclear reactors for both electricity and hydrogen production could significantly contribute to a more sustainable and secure energy future. Additionally, the integration of hydrogen production with water desalination offers a promising avenue for addressing water scarcity challenges, further amplifying the environmental and societal benefits of pink hydrogen production.

The development of blue and pink hydrogen production methods exemplifies the innovative strategies being pursued to meet the world's growing energy demands while minimizing environmental impacts. These methods, rooted in nuclear technology, pave the way for a future where energy production is both sustainable and versatile.

6.2.4 Gray Hydrogen

6.2.4.1 Overview of Gray Hydrogen

Gray hydrogen (GH_2) currently dominates the hydrogen market, primarily produced through natural gas steam reforming. This method, however, does not capture the greenhouse gases generated during the process. GH_2 is extensively used in the decarbonization of industrial sectors that consume significant electricity, such as in the production of metallurgical products and cement. It is essentially similar to blue hydrogen but lacks the CCS aspect. Representing about 95% of the market, gray hydrogen is a significant yet environmentally contentious player.

6.2.4.2 Costs and Environmental Implications

As of 2020, the estimated production costs range from $1 to 1.80/kg for gray and blue hydrogen and $2.50–6.80/kg for green hydrogen. Annually, approximately 94 million tons of gray hydrogen are produced globally using fossil fuels, marking it as a substantial source of greenhouse gas emissions.

6.2.5 Turquoise Hydrogen

6.2.5.1 Innovative Production of Turquoise Hydrogen

Turquoise hydrogen, identified for its zero-greenhouse gas emission profile, is produced through methane pyrolysis. This technology operates at around 1065 °C and allows for hydrogen extraction from natural gas while sequestering solid carbon. Unlike traditional methods, this process directly generates high-quality hydrogen from natural gas and separates the solid carbon, which is neither released into the atmosphere nor contributes to pollution. Instead, it can be utilized for industrial purposes or safely stored in landfills.

This one-step process not only produces hydrogen in high volumes but also does so at a lower cost compared to steam reforming.

6.2.6 Yellow Hydrogen

6.2.6.1 Thermal Solar Production of Yellow Hydrogen

Yellow hydrogen involves the dissociation of water into hydrogen and oxygen at very high temperatures. The addition of catalysts lowers the required temperature, making the process more feasible. Concentrating solar power is utilized to heat water, achieving the necessary temperatures of 800–1,200 °C. This method is exemplified by the 100-kW pilot plant at the Plataforma Solar de Almería in Spain, operational since 2008.

This technology demonstrates the potential for scaling up hydrogen production. To increase the production capacity, there would be a need for more reactor units and an expanded field of sun-tracking mirrors. Such scalability could pave the way for widespread implementation, combining hydrogen production with electricity generation and contributing to a sustainable energy future.

6.3 Current Industrial Hydrogen Production

Hydrogen has been a cornerstone of industrial processes for over a century. Despite being the most abundant element in the universe, hydrogen is not readily available in its free form on Earth, necessitating extraction from its compounds. This element's unique properties make it challenging to produce in large quantities.

6.3.1 Global Production and Market Value

As of 2020, approximately 87 million tons of hydrogen were produced worldwide, serving various industrial needs. Key applications include oil refining, ammonia production via the Haber process, and methanol production through the reduction of carbon monoxide. The global hydrogen generation market, valued at US$135.94 billion in 2021, is projected to expand to US$219.2 billion by 2030. This growth trajectory indicates a compound annual growth rate (CAGR) of 5.4% from 2021 to 2030.

6.3.2 Industrial Methods of Hydrogen Production

Various industrial methods exist for producing hydrogen, either in gaseous or in liquid form. These techniques range from well-established commercial processes to emerging technologies under development. Each method has its specific application, efficiency, and environmental impact, contributing to the diverse landscape of hydrogen production. This diversity is key to meeting the growing global demand for hydrogen in various sectors, from energy to manufacturing.

6.3.3 Electrolysis of Water

6.3.3.1 Historical Development and Fundamentals

The concept of electrolysis, pivotal in hydrogen production, was pioneered by British scientist Sir William Robert Grove in 1839. His groundbreaking experiments demonstrated the feasibility of splitting water into hydrogen and oxygen using electricity. Additionally, he explored the reverse reaction, where combining oxygen and hydrogen could generate electricity, laying the groundwork for modern fuel cell technology.

6.3.3.2 Electrolysis Techniques in Hydrogen Production

Various methods exist for producing hydrogen via electrolysis:

- *High-Pressure Electrolysis*: Works under elevated pressures, increasing efficiency.
- *Low-Pressure Electrolysis of Water*: Standard method, operating at lower pressures.
- *Advanced Electrochemical Processes*: Including high-temperature electrolysis (HTE) and carbon-assisted electrolysis, these are emerging as innovative alternatives.

Despite advancements, current optimal processes for water electrolysis achieve an electrical efficiency between 70% and 80%. Consequently, producing 1 kg of hydrogen (with a specific energy content of 143 MJ/kg, approximately 40 kWh/kg) requires about 50–55 kWh of electricity.

6.3.3.3 Electrolysis Reactions

The process of electrolysis involves splitting of water into hydrogen and oxygen gases through electric current. The cathode captures electrons from the hydronium ions to produce hydrogen gas, while at the anode, water molecules lose electrons, forming oxygen gas. The reactions can be outlined as follows:

- Oxidation at Anode: $6\ H_2O\ (l) \rightarrow O_2\ (g) + 4H_3O^+\ (aq) + 4\ e^-$
- Reduction at Cathode: $4H_3O^+\ (aq) + 4\ e^- \rightarrow 2\ H_2\ (g) + 4\ H_2O\ (l)$
- Overall Reaction: $2\ H_2O\ (l) \rightarrow 2\ H_2\ (g) + O_2\ (g)$

Several types of electrolyzers are employed, depending on the electrolyte used:

1) *Alkaline Electrolyzers*: Use a liquid electrolyte, usually a 25% potassium hydroxide solution.
2) *Polymer Electrolyte Membrane (PEM) Electrolyzers*: Feature polymer membranes as electrolytes, distinguished by their proton exchange membranes, which facilitate gas separation.
3) *Solid Oxide Electrolyzers*: Employ ceramic materials as both the electrolyte and separator membrane, effectively isolating hydrogen and oxygen.
4) *Seawater Electrolyzers*: Utilize seawater, an abundant resource, but face challenges due to the production of toxic chlorine gas from high sodium chloride content.
5) *Solar-Powered Electrolyzers*: Leverage solar energy to power the electrolyzer, promoting renewable energy use and reducing reliance on fossil fuels.

6.3.3.4 High-Pressure Electrolysis

High-pressure electrolysis, which outputs hydrogen at 120–200 bar, offers enhanced system efficiency. By pressurizing hydrogen within the electrolyzer through chemical compression, it eliminates the need for external compressors, thereby reducing energy consumption. Europe's largest hydrogen production plant in Kokkola, Finland, exemplifies the application of high-pressure alkaline electrolysis, with an expected annual capacity of ~1,500 tons.

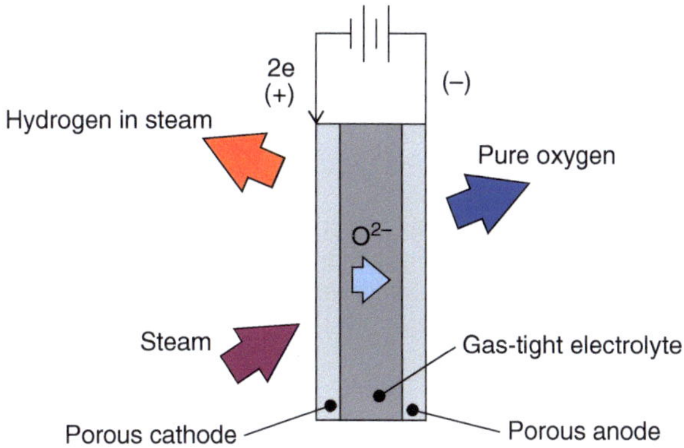

Fig. 6.5 High-Temperature Electrolysis. *Source:* Alexpreeves/Wikimedia Commons/CC BY-SA 4.0

6.3.3.5 High-Temperature Electrolysis

HTE provides a method to produce hydrogen using both heat and electricity. By supplying energy partially in the form of heat, HTE reduces the amount of energy conversion needed, potentially lowering the energy demand per kilogram of hydrogen produced. This process has shown promise in laboratory settings, achieving 108 MJ (thermal) per kilogram of hydrogen. However, its commercial viability remains unproven, and the hydrogen produced is typically of commercial grade, not suitable for fuel cells (Fig. 6.5).

6.3.3.6 The Evolution of Electrolysis Technology

The evolution of electrolysis technology reflects a continual quest for efficiency and environmental sustainability in hydrogen production. From its early inception to modern advancements, electrolysis remains a critical component in the hydrogen economy. As technology progresses, new methods like solar-powered electrolyzer and HTE offer avenues for greener and more efficient hydrogen production. The integration of these technologies into existing energy systems presents opportunities for significant advancements in renewable energy utilization and a reduction in carbon emissions.

6.3.4 Water Splitting Using Solar Energy

6.3.4.1 Thermochemical Cycles for Water Splitting

Various thermochemical cycles, including the iron oxide cycle, cerium (IV) oxide–cerium (III) oxide cycle, zinc–zinc oxide cycle, sulfur–iodine cycle, copper–chlorine cycle, hybrid sulfur cycle, and aluminum–aluminum oxide cycle, are currently in research and testing phases. These methods aim to produce hydrogen and oxygen from water and heat without using electricity. The efficiency of

1. Photoelectrochemical water splitting

- The nanotech-based metal oxide photoelectrochemical materials could lower the cost of hydrogen production somewhere between a factor of 4 and 10
- The serious limitation is susceptibility to corrosion of the nanomaterials that to be immersed in water. Research is in progress to find more efficient, lower cost components and systems that are durable and stable against corrosion in water surroundings

Fig. 6.6 Photoelectrochemical Water Splitting

these processes can surpass HTE, typically ranging from 35% to 49% in terms of lower heating value (LHV) efficiency. However, thermochemical production of hydrogen using chemical energy from coal or natural gas is generally overlooked due to the higher efficiency of direct chemical pathways (Fig. 6.6).

While several of these thermochemical hydrogen production processes have been demonstrated in laboratory settings, none have yet reached production levels.

6.3.4.2 Photoelectrochemical Water Splitting

Innovations in photoelectrochemical water splitting were pioneered by Energy Conversion Devices in 1983. They patented the first multijunction high-efficiency photoelectrochemical system, which has led to the development of what is now referred to as an "artificial leaf" or "wireless solar water splitting." This technology utilizes a low-cost thin-film amorphous silicon multijunction sheet immersed directly in water.

6.3.4.3 Photocatalyst Development by Panasonic Corp.

Panasonic Corp. has developed a significant advancement in photocatalysis, creating a niobium nitride-based photocatalyst capable of absorbing 57% of sunlight. This technology supports the breakdown of water to produce hydrogen gas (Figs. 6.7 and 6.8).

6.3.4.4 Wind Energy and Hydrogen Production

Currently, there are no large-scale projects initiated for directly converting wind energy into hydrogen. However, the potential of using wind energy for hydrogen storage in limited applications is a real possibility. This approach involves converting wind energy into electrical energy, which can then be used for hydrogen production through electrolysis. The concept opens avenues for integrating renewable wind energy into the hydrogen economy, particularly in regions with abundant wind resources.

2. Solar thermal water splitting

- Highly concentrated sunlight can be used to produce the high temperatures required to split methane into hydrogen and carbon
- Concentrated solar energy can also be used to generate temperatures of several hundred to over 2,000 degrees at which thermochemical reaction cycles can be used to produce hydrogen
- Such high-temperature, high-flux solar driven thermochemical processes offer a novel approach for the environmentally benign production of hydrogen
- Very high reaction rates at these elevated temperatures give rise to very fast reaction rates that enhance the production rates significantly and more than compensate for the intermittent nature of the solar resource

Fig. 6.7 Solar Thermal Water Splitting

3. Photobiological water splitting

- Certain photosynthetic microbes produce hydrogen from water in their metabolic activities using light energy
- Photobiological technology holds great promise, but because oxygen is produced along with the hydrogen, the technology must overcome the limitation of oxygen sensitivity of the hydrogen-evolving enzyme systems
- Researchers are addressing this issue by screening for naturally occurring organisms that are more tolerant of oxygen, and by creating new genetic forms of the organisms that can sustain hydrogen production in the presence of oxygen
- A new system is also being developed that uses a metabolic switch (sulfur deprivation) to cycle algal cells between a photosynthetic growth phase and a hydrogen production phase

Fig. 6.8 Photobiological Water Splitting

6.3.5 Biomass Gasification with Bacteria

Microbial biomass conversion processes utilize the ability of microorganisms to consume and digest biomass, subsequently releasing hydrogen. In fermentation-based systems, microorganisms, such as bacteria from the Firmicutes phylum or Clostridium genus, break down organic substances. These substances range from refined sugar to raw biomass sources like corn stover, and even wastewater, producing hydrogen in a manner similar to

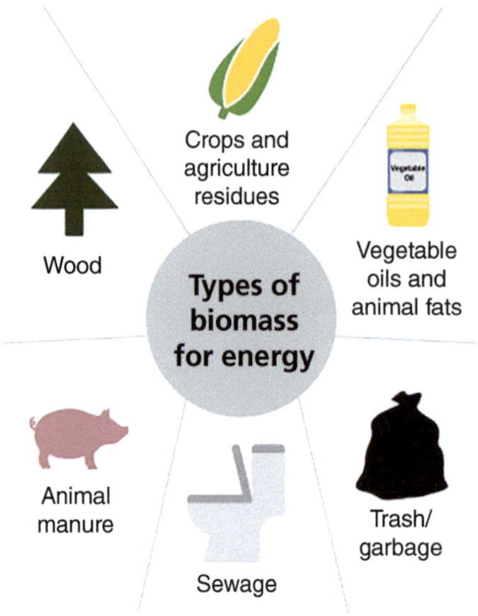

Fig. 6.9 Biomass Sources for Fuel Production. *Source:* U.S. Energy Information Administration

anaerobic conversion. Termed "dark fermentation," these methods do not require light, enabling constant hydrogen production from organic compounds regardless of the time of day (Fig. 6.9).

Dark fermentation reactions, due to their independence from light energy, offer the advantage of continuous hydrogen production. In contrast, photofermentation requires light and operates differently from dark fermentation. Electrohydrogenesis, another method used in microbial fuel cells, produces hydrogen from organic matter, such as sewage or solid waste, while applying a voltage of 0.2–0.8 V.

Current research focuses on enhancing the rate and reducing the production time in fermentation systems. An emerging area is the development of microbial electrolysis cells (MECs). MECs exploit the energy and protons generated by microbes breaking down organic matter, combined with a small electric current, to produce hydrogen.

6.3.6 Hydrogen as a By-product of Other Chemical Processes

6.3.6.1 Ammonia Dissociation

Ammonia dissociation, a process that involves breaking down ammonia (NH_3) into hydrogen and nitrogen, occurs at approximately 1,800 °F with the assistance of a catalyst. The resulting product, a mixture of 75% hydrogen and 25% mononuclear nitrogen (N rather than N_2), finds direct application in the metal industry for processes such as brazing or bright annealing.

6.3.6.2 Hydrogen in Industrial Production

In the industrial production of chlorine and caustic soda through electrolysis, a significant amount of hydrogen is generated as a byproduct. Similarly, the gas produced from coke ovens in steel manufacturing resembles Syngas, containing about 60% hydrogen by volume. This hydrogen can be economically extracted from the coke oven gas, adding another dimension to industrial hydrogen production.

6.3.7 Municipal Solid Waste (MSW) Utilization for Hydrogen Production

MSW and biomass are significant sources of anthropogenic methane emissions, primarily through landfill gas (LFG) production. The generation of LFG involves several mechanisms:

- *Bacterial Decomposition*: Breakdown of organic matter by bacteria.
- *Volatilization*: Conversion of solid waste components into gaseous form.
- *Chemical Reactions*: Interaction of waste materials leading to gas production (Fig. 6.10).

Fig. 6.10 Landfill Preparation Stage. *Source:* Marcus Clackson/Getty Images

Fig. 6.11 Santa Marta Landfill Gas Capture Facility in Talagante, part of the Santiago Metropolitan Region in Chile. *Source:* https://www.goclimate.com/blog/santa-marta-landfill-gas-recovery/

6.3.7.1 Composition and Utilization of LFG

LFG typically comprises about 50% methane, 45% carbon dioxide, 3% nitrogen, 1% oxygen, and 1% nonmethane organic vapors. It can be harnessed to generate electricity and heat, commonly using internal combustion engines such as diesel engines or microturbines. Notably, LFG contains approximately half the heating value of compressed natural gas (CNG), making it an excellent source for hydrogen production. This attribute positions LFG as a viable, alternative feedstock for hydrogen production, contributing to sustainable energy solutions and waste management practices (Fig. 6.11).

6.3.7.2 Potential for Hydrogen Production

The utilization of LFG for hydrogen production presents an innovative approach to waste-to-energy conversion. By tapping into the energy potential of MSW, this method not only addresses waste management challenges but also adds to the renewable sources of hydrogen, aligning with global efforts toward environmental sustainability and energy diversification (Fig. 6.12).

6.4 Traditional Hydrogen Production Methods

The majority of the hydrogen market is dominated by gray hydrogen, primarily produced through natural gas steam reforming without capturing the resultant greenhouse gases. This method, while economically viable, poses significant environmental challenges due to its carbon footprint. The chapter also explores the

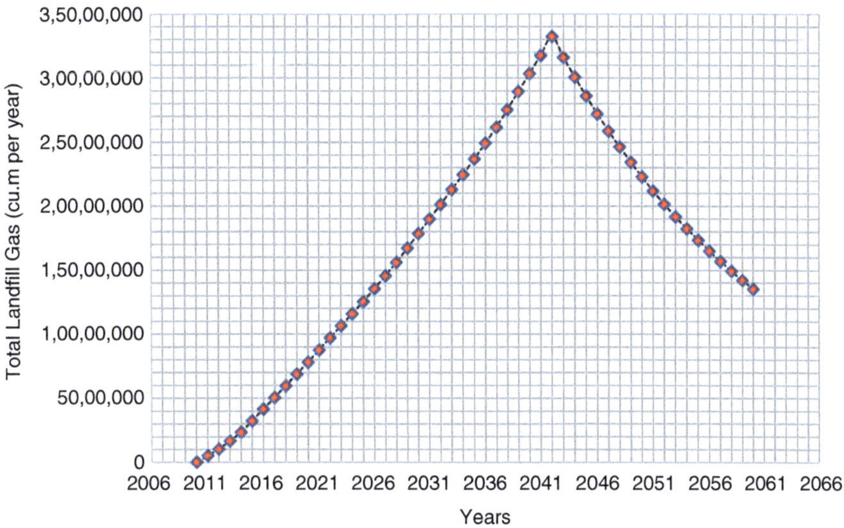

Fig. 6.12 Total Landfill Gas emissions in cu. m per year over 2010-2060 from the Tirupati municipal solid waste dumpsite near Tirupati, India. *Source:* https://www.sciencedirect.com/science/article/pii/S2215016122002485

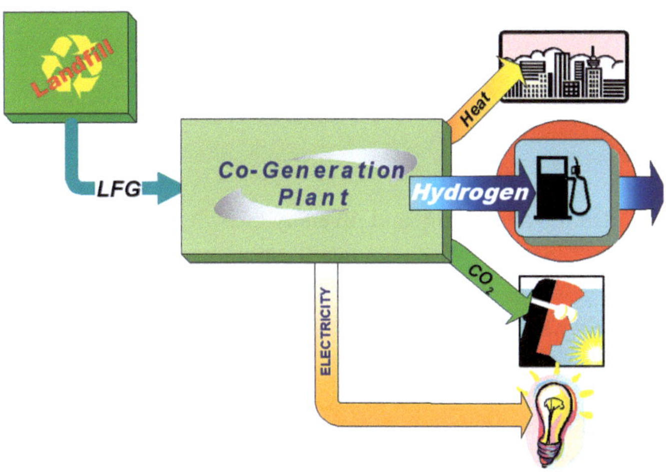

Fig. 6.13 LFG Co-generation Plant Integrated with Hydrogen Production

role of ammonia dissociation and other industrial processes, like chlorine and caustic soda production, in generating hydrogen as a by-product. These methods underscore the interconnectivity of hydrogen production with other industrial processes and their potential environmental impacts (Fig. 6.13).

6.4.1 Renewable Energy Sources in Hydrogen Production

A significant focus is placed on harnessing renewable energy sources for hydrogen production. The potential of solar energy, particularly through photoelectrochemical water splitting and the use of photocatalysts like niobium nitride, is discussed. These innovative approaches demonstrate the feasibility of using solar energy directly in hydrogen production, reducing reliance on fossil fuels.

The chapter also examines the use of wind energy for hydrogen storage and its potential applications. While direct conversion of wind energy into hydrogen is not extensively developed, the concept of using wind energy in hydrogen storage opens new avenues for integrating renewable energy sources into the hydrogen economy.

Emerging technologies, such as biomass gasification with bacteria, offer novel approaches to hydrogen production. These methods leverage the capability of microorganisms to break down organic matter and release hydrogen, illustrating the potential of biological processes in renewable hydrogen production.

Another innovative method discussed is MSW utilization. By converting MSW into LFG, which contains a significant proportion of methane, hydrogen production can be achieved while addressing waste management challenges. This approach not only contributes to sustainable energy solutions but also aligns with global efforts toward environmental sustainability and resource efficiency.

6.5 Conclusion

This chapter encapsulates the complexity and diversity of hydrogen production methods. From traditional methods with significant environmental impacts to innovative approaches leveraging renewable energy sources and waste materials, the landscape of hydrogen production is evolving rapidly. Understanding these various methodologies is crucial for developing sustainable, efficient, and environmentally friendly hydrogen production systems. As we advance in technology and innovation, these methods are poised to play an increasingly important role in the global transition to sustainable energy solutions.

7

Hydrogen Applications

7.1 Current Industrial Applications

Technological advancements have historically pivoted around efficient energy utilization, transitioning from coal to nuclear, and more recently, to renewable sources. While it may be premature to herald the onset of a hydrogen revolution as the dawn of a new human era, the potential for a significant increase in hydrogen use is undeniable. Hydrogen energetics has, for many years, been an integral component of numerous applications, serving a diverse array of business consumers. Below, we detail some of the primary uses of hydrogen in various industries.

Currently, hydrogen is primarily utilized as an industrial feedstock, notably in ammonia and methanol production, petroleum refining, and as rocket fuel. In the United States, the industry consumes nearly all hydrogen for petroleum refining, metal treatment, fertilizer production, and food processing, underscoring its integral role in industrial, space, and military applications.

7.1.1 Ammonia Synthesis

The cornerstone of industrial ammonia production remains the Haber–Bosch process, which synthesizes ammonia (NH_3) through the combination of hydrogen and nitrogen in the presence of a metal catalyst under high pressure and temperature. Ammonia serves a dual purpose: it is a precursor to ammonium nitrate, a widely used fertilizer, and it is a constituent in many household cleaning products. Following oil refineries, ammonia synthesis is the largest consumer of hydrogen.

Most ammonia plants generate hydrogen via steam methane reforming (SMR) technology. Notably, the world's largest single-train ammonia plant, located in Al-Jubail, Saudi Arabia, boasts a production capacity of 1,300 tons/day.

Hydrogen Energetics: A Future Clean Energy Carrier, First Edition. Roman J. Press.

7.1.2 Food and Beverage Industry

The food and beverage industry consumes large quantities of hydrogen gas, primarily in the process of hydrogenation, where unsaturated fats are converted into saturated oils and fats. This hydrogenation process is critical for producing hydrogenated vegetable oils, such as margarine and shortening, from unsaturated vegetable oils (fatty acids), thereby obtaining solid fats.

Hydrogenation is conducted in batches within a heated tank. The process begins by pumping the oil feed (e.g., sunflower seed or olive oil) into a heated pressure vessel. A vacuum is applied to prevent oxidation as the temperature is raised between 140 and 250 °C, with continuous stirring to maintain even heating. Nickel catalyst solids, premixed with a small amount of oil, are introduced, followed by the infusion of hydrogen gas, which increases the pressure to 2.7–4 bar. Given the exothermic nature of the hydrogenation reaction, external heating is ceased, and cooling is applied to maintain a temperature range of 70–80 °C. After 40–60 minutes, the hydrogenated oil mixture is extracted as a slurry, and catalyst solids are separated using filters. Cooling the mixture to room temperature allows the hydrogenated oil to solidify.

7.1.3 Electronics Manufacturing

Hydrogen is indispensable in the field of electronics manufacturing, serving key roles in the production of semiconductors, displays, LEDs, and photovoltaic technologies. The fabrication of these miniature electronic devices entails a series of complex semiconductor manufacturing processes, including oxidation, photolithography, diffusion, epitaxy, and aluminum evaporation. The creation of integrated circuits involves two primary stages: wafer manufacturing – encompassing crystal growth, slicing, edge grinding, polishing, and wrapping – and chip assembly. Hydrogen is a vital component in these processes, especially for the deposition of semiconducting layers on integrated circuits and during silicon deposition or crystal growing.

- *Annealing*: This process involves heating silicon wafers to temperatures above 1,000 °C to repair or anneal their crystal structure. Hydrogen ensures uniform heat distribution, assists in reconstructing the crystal structure on the surface layers, and aids in oxide removal.
- *Epitaxy*: Utilized as a reducing agent, hydrogen is crucial in epitaxial processes for depositing new crystalline films (see Fig. 7.1).
- *Deposition*: Hydrogen is directly incorporated into thin films to reduce crystallinity, enhancing their electrical insulating properties, notably in silicon thin films.
- *Stabilizing*: Additionally, hydrogen extends the shelf life of significant electronic chemicals, underscoring its importance in electronics manufacturing.

Fig. 7.1 Electronic Assembly Board. *Source:* Raimundas/Adobe Stock Photos

7.1.4 Pharmaceutical Industry

Hydrogen's applications extend into the pharmaceutical industry, where it is used in the synthesis of vitamins and other products. Hydrogen peroxide (H_2O_2), a potent oxidizing agent, is widely employed as a sterilizing agent in medical facilities for cleaning wounds and other tissue damages. Beyond its medical applications, H_2O_2 is used for bleaching hair, whitening teeth, and stain removal from clothing. Its utility in research, particularly for evaluating the antioxidant potential of enzymes like catalase, highlights its versatility.

7.1.5 Glass, Cement, and Lime Production

In the production of glass, cement, and lime, hydrogen, when combined with nitrogen, serves as a reductive agent in processes such as the float glass technique, preventing oxidation of the tin bath. It is also a key component of the oxy-hydrogen flame used for the heat treatment of glass preforms, demonstrating its wide-ranging utility in industrial manufacturing.

7.1.6 Polymers

Polymers, essential to modern life, are large molecules formed by bonding smaller units called monomers. The term "polymer" derives from Greek words meaning "many parts," highlighting the diverse and complex structure of these materials. Polymers encompass a wide array of substances, including plastics, elastomers, fibers, adhesives, and surface coatings, distinguished by their molecular size and structure. They are categorized into natural, synthetic, or industrial polymers, with further classification based on their physical properties (thermoplastic and thermosetting resins) and polymerization processes (chain-growth or step-growth reactions). Hydrogen is pivotal in the chemical industry, contributing to the synthesis of key materials such as high-density polyethylene, polypropylene, and in the hydrogenation of unsaturated oils to produce food-grade fats.

7.1.7 Metal Industry

The metal industry utilizes hydrogen for its reductive properties, employing it to reduce metal oxides and prevent oxidation during the heat treatment of metals and alloys. This process is crucial for refining metals, enhancing their quality and durability. Hydrogen, mixed with inert gases, creates a reductive atmosphere, optimizing the heat treatment process for steel and other metals, ensuring the integrity and performance of the final products.

7.1.8 Metallic Ore Reduction

Hydrogen's role extends to metallic ore reduction, where it is used to extract valuable metals such as tungsten from ores like wolframite, scheelite, and ferberite. Its application in extracting copper from copper oxides (tenorite and paramelaconite) highlights its versatility. The concept of direct reduction of iron (DRI) using hydrogen represents a promising area for reducing carbon emissions in steel production, although it has not yet achieved widespread commercial use. This process primarily yields water vapor and nitrogen, significantly lowering CO_2 output compared to traditional methods.

7.1.9 Oil and Gas

In the oil and gas industry, hydrogen technology is advancing rapidly, especially in refining processes. Hydrogen is crucial for desulfurizing fuel oil and gasoline, removing harmful sulfur compounds, and enhancing fuel quality. Hydrocracking, a process that breaks down heavy, unsaturated hydrocarbons into lighter, more stable compounds using hydrogen and a catalyst, is essential for producing high-quality distillates such as diesel and petrol. This application underscores hydrogen's significant role in improving efficiency and sustainability in fuel production.

7.1.10 Methanol Production

Methanol, a potential fossil-fuel substitute, is traditionally synthesized from synthesis gas – a mixture of carbon monoxide and hydrogen – in a fixed-bed reactor. This reactor uses alumina pellets coated with copper and zinc oxides as a catalyst.

An emerging method involves the direct combination of hydrogen and carbon dioxide, a reaction gaining attention for its potential to convert atmospheric CO_2 into usable fuel. The key challenge lies in achieving thermodynamic efficiency, where the energy contained in the produced methanol exceeds the energy expended in its production. Recent research has focused on identifying an effective catalyst to ensure high selectivity and efficiency in methanol production. A breakthrough in the United States revealed that nanoparticles of a palladium and copper catalyst, dispersed on a porous support material to maximize surface area, yield the highest efficiency. Remarkably, a catalyst pellet the size of a walnut has an internal surface area equivalent to a football field.

In this innovative process, hydrogen and carbon dioxide are introduced into a reactor vessel filled with the catalyst and heated to 180–250 °C. The process achieves a maximum CO_2-to-methanol conversion rate of about 24%, with unconverted carbon dioxide and hydrogen being recycled back into the system. The process's overall thermodynamic efficiency remains unspecified.

7.1.11 Automotive and Transportation

Hydrogen fuels a new frontier in automotive and transportation, promising to eliminate carbon emissions in internal combustion engines (ICEs). Its application in fuel cell vehicles (FCVs) highlights a shift toward clean energy production, a topic explored in detail in Chapter 6. This development underscores hydrogen's pivotal role in driving sustainable automotive solutions.

7.1.12 Space-Aviation

Hydrogen's role in space exploration cannot be overstated. Primarily utilized in its gaseous state, hydrogen is also stored and transported as a liquid for space endeavors. Since the 1950s, the National Aeronautics and Space Administration (NASA) has leveraged liquid hydrogen as a rocket fuel, marking it as a pioneer in using hydrogen fuel cells (FCs) for powering spacecraft electrical systems. In the space program, liquid hydrogen is a preferred fuel, combusted with oxygen or fluorine, and it serves as a propellant for nuclear-powered rockets and space vehicles, highlighting its critical role in propelling humanity's exploration of the final frontier (see Fig. 7.2).

7.1.13 Hydrogen in Welding, Cutting, and Coating

In the industrial sphere, hydrogen's applications extend to the heat treatment of metals, including the annealing of stainless steel and magnetic steel alloys, sintering, copper brazing, and precision cutting. Atomic hydrogen welding (AHW), or plasma welding, utilizes an arc between tungsten electrodes in a hydrogen

Fig. 7.2 NASA Liquid Storage. *Source:* NASA/Public Domain

atmosphere to weld refractory metals and tungsten effectively. This process underscores hydrogen's versatility and essential role in advanced manufacturing and metalworking.

7.1.14 Weather Balloons

Meteorologists rely on hydrogen for high-altitude weather balloons due to its lighter-than-air quality. This application takes advantage of hydrogen's buoyancy, enabling weather instruments to ascend to the upper atmosphere for crucial data collection and weather observation, playing a vital role in forecasting and climate research.

7.1.15 Hydrogen as a Coolant

Hydrogen's excellent thermal properties are harnessed in modern large electrical generators, where it serves as a rotor coolant at around four bars pressure. Its unmatched thermal conductivity efficiently draws heat away from the stator, offering significant advantages:

- Its low density reduces windage loss by about 10% compared to air, enhancing operational efficiency.
- Superior thermal conductivity facilitates the use of smaller cooling systems.

- The highly specific heat capacity enables effective heat management.
- Being cleaner than air, hydrogen minimizes the risk of reducing the electrical resistance of bushings, ensuring the longevity and reliability of electrical systems.

7.1.16 Searching Gas

Hydrogen serves as an effective leak detection medium in manufacturing facilities, favored for its minimal environmental impact compared to $CClF_3$-based gases historically used. It can be employed alone or combined with other elements, acting as a crucial reducing agent in redox reactions. For example, in the production of plate glass, hydrogen prevents the formation of stannous oxide (SnO) in the float bath, showcasing its versatility in manufacturing processes.

7.1.17 Chemical Analysis

In the realm of chemical analysis, hydrogen's applications are varied and essential. It fuels atomic absorption spectroscopy, generating heat to produce neutral atoms for analysis. Additionally, hydrogen is utilized as a carrier phase in gas chromatography, facilitating the separation of volatile substances. Its role extends to the study of superconductivity, where liquid hydrogen, due to its melting point near absolute zero, is instrumental.

7.1.18 Isotope Applications

Hydrogen isotopes play pivotal roles in various sectors. Deuterium, for instance, is used as a moderator in nuclear reactors to slow down neutrons and as a tracer in scientific research. Tritium, another isotope, finds applications in the production of hydrogen bombs, luminous paints, and as a tracer, highlighting the element's significance in both energy production and scientific studies.

7.1.19 Burning Hydrogen for Electricity Generation

The momentum to utilize hydrogen as a fuel for power generation is increasing in the United States. Power plants are beginning to operate on natural gas–hydrogen mixtures in gas turbines. For instance, the Long Ridge Energy Generation Project in Ohio plans to run a gas turbine on a 95% natural gas and 5% hydrogen blend, aiming to eventually transition to 100% green hydrogen from renewable sources. Similarly, the Intermountain Power Agency in Utah is converting a coal-fired plant to a gas-fired facility that will initially incorporate up to 30% hydrogen, with aspirations to shift entirely to 100% green hydrogen, marking a significant step toward sustainable energy production.

Hydrogen FCs, a cornerstone of modern clean energy technologies, produce electricity by catalyzing a reaction between hydrogen and oxygen atoms. This process, akin to a battery's electrochemical reaction, generates electricity, water, and a small amount of heat. The versatility of FCs allows for their application across a broad spectrum, from portable devices like laptops and cell phones to significant military uses. Larger scale implementations include providing electricity to power grids, offering backup power for buildings, and serving remote locations not connected to the electric power infrastructure.

As of late October 2021, the United States had approximately 166 operational FC electric power generators across 113 facilities, summing up to about 260 MW of electric generation capacity. The Bridgeport Fuel Cell in Connecticut stands out with about 16 MW of generation capacity, making it the largest single unit. Following closely are two 6 MW units, including one at the Red Lion Energy Center in Delaware, which, combined with 5 smaller units, achieves a total generation capacity of 25 MW. While most of these operational FCs utilize pipeline natural gas as the hydrogen source, alternative sources like landfill gas and biogas from wastewater treatment are also in use, highlighting the flexibility in hydrogen sourcing.

The innovation in hydrogen energy extends to projects like the San Diego Gas and Electric's power-to-gas-to-power initiative. This project uniquely integrates the electric grid to produce hydrogen via electrolysis, subsequently using this hydrogen in FCs to generate electricity. Such initiatives underscore the evolving landscape of hydrogen use in energy production, emphasizing its potential to serve as a sustainable energy carrier.

The allure of hydrogen as a substitute for natural gas in heating and electricity generation is rooted in its high calorific value and the environmentally benign nature of its combustion products, primarily water. However, realizing these benefits necessitates overcoming several challenges associated with hydrogen's physical and chemical properties. These challenges include managing its high flame temperature and speed, which could lead to increased NO_x emissions and unstable combustion, addressing compression difficulties due to its low molecular mass, navigating the complexities of large-scale storage given its lower energy density compared to natural gas, and mitigating risks associated with its low ignition energy.

Historically, hydrogen's production traces back to 1650 with Théodore de Mayerne's experiment, marking the beginning of hydrogen's journey from a curiosity to a critical component of the modern energy landscape. Notable milestones in hydrogen technology include the development of hydrogenation in 1897, the Haber process for synthesizing ammonia in 1910, and the advent of hydrocracking in 1920. These advancements laid the groundwork for hydrogen's diverse applications today, from FCs to industrial processes, underscoring its role as a premium product produced with purities up to 99.999%.

Today, the interest in hydrogen FCs for electricity generation is part of a broader push toward sustainable energy solutions. This interest is driven by the urgent need to transition away from fossil fuels and mitigates the impacts of climate change. As the technology for hydrogen production and utilization continues to evolve, the potential for hydrogen to revolutionize energy systems worldwide grows, positioning it as a key player in the quest for clean, sustainable energy.

7.2 FC-Specific Applications

7.2.1 Stationary Power Production

The market for stationary power generation systems utilizing FCs encompasses a diverse array of applications, including remote sites, telecommunications infrastructure, commercial and residential buildings, backup power sources, and auxiliary power units (APUs). These systems vary in size, typically ranging from 3 to 350 kW.

Stationary power production employs four primary FC technologies:

- Molten carbonate fuel cells (MCFCs) are distinguished by their high efficiency, surpassing other FC types, and are not as affected by the high-temperature material challenges that impact solid oxide fuel cell (SOFC) technology. MCFCs are particularly suitable for large-scale power plants fueled by natural gas or coal and are applicable in electrical utilities, industrial settings, and military operations. Operating at temperatures above 650 °C, MCFCs utilize a molten carbonate salt mixture within a porous, inert ceramic matrix as an electrolyte. This high operational temperature allows for the use of cost-effective, nonprecious metal catalysts. Efficiencies can approach 60%, significantly higher than the 37–42% typical of phosphoric acid fuel cell (PAFC) plants. When combined with cogeneration to utilize waste heat, overall efficiencies can reach up to 85%. However, the durability of MCFCs is a concern; the corrosive nature of the electrolyte and the high temperatures accelerate wear and corrosion, leading to reduced cell lifespan.
- PAFCs utilize liquid phosphoric acid as the electrolyte and carbon paper electrodes coated with a platinum catalyst. A notable advantage of PAFCs is their tolerance to impurities in the hydrogen feed. The solidification point of the phosphoric acid at 40 °C can complicate start-up, but within the operational range of 150–200 °C, the system can repurpose expelled water as steam for heating purposes. PAFCs are prevalent in stationary applications, achieving combined heat and power efficiencies around 80%. UTC Fuel Cells is a leading PAFC producer, with numerous "PureCell" 200 kW units globally. However, PAFCs require periodic replacement of the fuel-cell stack and reformer, typically every five years, prompting ongoing development efforts focused on enhancing durability and reducing costs.

- SOFCs are under development by several companies, including GE Energy, which is working on a multi-megawatt system integrating an SOFC with a gas turbine for coal-based power generation. The current trend is to reduce the operational temperature from 1,000 to 800 °C to mitigate the challenges of building large-scale units, such as limited test data availability and the high cost of resources.
- Proton exchange membrane fuel cells (PEMFCs) have seen limited use in stationary applications due to the requirement for pure hydrogen. Nonetheless, advancements are being made in PEMFC systems that can process various fuels into hydrogen, including natural gas, propane, kerosene, JP-8, biodiesel, and alcohol-based fuels, for power outputs around 50 kW.

In 2005, Verizon Communications commissioned the largest FC installation to date, a 1.4 MW plant that also recovers 30,000 kWh of heat for building use. This system offsets approximately 50,000 tons of carbon dioxide annually compared to a fossil fuel plant of similar size.

Fuel cell and gas turbine hybridization involves the integration of fuel cells with gas turbines (FC-GT) to enhance the efficiency of large stationary power stations. By recuperating high-temperature waste heat, the FC-GT hybrid cycle can achieve electrical efficiencies above 70%. For instance, FuelCell Energy's DFC/T technology combines a direct fuel cell (DFC) with a gas turbine operating at ambient pressure, with the FC contributing the majority of the power output. Such systems are projected to reach efficiencies of 80%.

FCs are increasingly being powered by alternative fuels, including coal gas, anaerobic digester gas, and gasified biomass, expanding the versatility and sustainability of this technology.

7.2.2 FC Transportation

A FC power plant is an electrochemical device that converts hydrogen gas energy into electrical power, which is then used for propulsion in FCVs. However, a FC alone is insufficient; it must be integrated with various subsystems and components to provide practical power for transportation (refer to Section 4.3.1).

Vehicle efficiency in FCVs is characterized by the conversion of hydrogen energy into electrical and then mechanical power. The electrical efficiency of an FCV can be expressed as the ratio of net electrical power output to the lower heating value (LHV) of hydrogen at the FC stack's input.

Fuel economy is a critical measure, especially as hydrogen production currently bears a higher cost than gasoline. Despite this, FCVs demonstrate superior fuel economy due to the higher efficiency of electrochemical conversion and reduced heat loss compared to ICEs. Fuel economy is measured in miles per gallon gasoline equivalent (MPGe), where the energy content of 1 US gallon of gasoline is equivalent to 1.09 kg of hydrogen.

Driving range is a pivotal design and performance parameter for FCVs, influencing the planning of hydrogen infrastructure and refueling stations. Presently, the range is constrained by the onboard hydrogen storage capacity, typically not exceeding 300 miles.

Emissions from FCVs are significantly lower than those from ICE vehicles, with the primary by-product being water. However, the production of hydrogen from fossil fuels can result in emissions of CO_2 and other harmful gases, underscoring the importance of renewable hydrogen sources for truly emission-free transportation.

Cost considerations are paramount as FCVs are currently produced in low volumes and at high costs. Mass production is anticipated to reduce costs, making FCVs competitive with conventional ICE vehicles. Advancements in HEV technology are expected to further drive down costs in the future.

7.2.3 Well-to-Wheel Analysis

General Motors Corporation, in collaboration with various entities, has conducted a "well-to-wheel" analysis (refer to Fig. 7.3), which is an exhaustive evaluation of the impact of greenhouse gas emissions and energy consumption related to transportation. This study, examining fuel sources, processing methods, and propulsion systems, was divided into two segments: (i) well-to-tank, focusing on fuel production and (ii) tank-to-wheel, concentrating on fuel utilization.

Well-to-Tank: This phase assesses the total energy consumption from the source to the tank. Petroleum-based fuels and compressed natural gas (CNG) exhibit the lowest energy losses in this stage. Conversely, the production of liquid hydrogen from centralized facilities and hydrogen through electrolysis, along with electricity and cellulosic ethanol, is associated with higher energy losses. In terms of

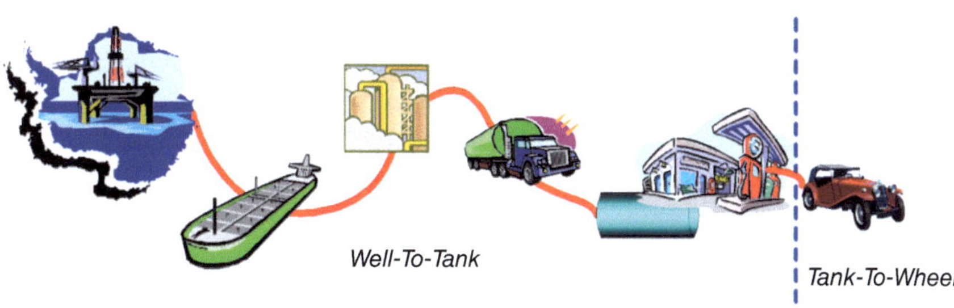

Fig. 7.3 Well-to-Wheel Diagram

greenhouse gas emissions, petroleum and natural-gas-based methanol production are more favorable, while hydrogen production from natural gas is less so due to significant emissions from the reformation process. Electrolysis for hydrogen production only minimizes emissions if the electricity used is sourced from nuclear or renewable energy.

Tank-to-Wheel: This analysis uses data from EPA-defined driving cycles, with vehicles such as the Chevrolet Silverado and the Opel Zafira serving as test models for North American and European studies, respectively. The study considered various fuels – diesel, ethanol, CNG, methanol, and hydrogen – in both ICEs and fuel cell systems (FCSs), as well as conventional and hybrid electric drives, all compliant with 2010 emissions standards.

FCSs showed superior fuel consumption rates over conventional technologies, with diesel hybrids demonstrating a significant reduction in greenhouse gas emissions when compared to standard gasoline engines. Ethanol-fueled vehicles and FC hybrids were found to have the lowest greenhouse gas emissions. A key takeaway was that the environmental benefits of FCVs are highly dependent on the hydrogen source, with the best outcomes arising from the use of renewable energy.

7.2.4 Practical Transportation Applications

Hydrogen can be utilized directly in ICEs. Ongoing developments in the United States and Europe are refining this technology. In the realm of FC transportation, oxygen is sourced from the air, but the production, distribution, and storage of hydrogen gas present complex challenges. Early FCV development explored on-board conversion of hydrocarbon-based fuels, with prototypes demonstrating fuel reformation and cleanup systems to produce hydrogen and remove harmful gases.

Commonly used fuels for on-board conversion include liquid methanol, CNG, and gasoline. Automakers like Daimler-Chrysler, Ford, and General Motors have constructed prototype vehicles with these systems. However, the complexity, cost, weight, and efficiency issues have limited the practicality of on-board fuel conversion (Fig. 7.4).

In recent years, energy companies such as Shell Hydrogen and Exxon-Mobil have committed to developing a global hydrogen infrastructure for vehicular use. FC buses, already operational in California, Europe, and Japan, are likely to be the public's first widespread encounter with FC technology, benefiting from centralized fueling and maintenance facilities.

7.2.5 Other Transport Applications

Aviation: Aircraft engines serve dual purposes: propulsion and providing comfortable conditions for passengers. While propulsion is still reliant on liquid fuels, the

Fig. 7.4 Hydrogen Bus. *Source:* scharfsinn86/Adobe Stock Photos

electrical load for heating and lighting can be supplied by FCs. This could reduce fuel consumption significantly, especially when aircraft are on the ground. Airbus and Boeing are developing APUs that could see on-board prototypes by the mid to late 2020s. These APUs, ranging from 200 to 600 kW, aim to be compact and lightweight to avoid encroaching on passenger space. On-board reforming of aviation fuel is an appealing prospect for simplifying the system and reducing costs and emissions (Fig. 7.5).

Off-road equipment: Small industrial vehicles like forklifts and tugs are also being considered for FC applications. Airports, with their repetitive vehicle use patterns, present opportunities for hydrogen fueling. FC-powered off-road equipment offers zero emissions, rapid refueling, and improved performance over battery-powered alternatives (Fig. 7.6).

Railways: In Europe, the Hydrogen Train Project aims to demonstrate a hydrogen FC-powered train, integrating a proton exchange membrane (PEM) stack with batteries for peak power. Japan is focusing on FC commuter trains, with demonstrations planned that will utilize a robust PEM fuel cell/battery hybrid system. In the United States, attention is also given to mining applications, where replacing diesel with hydrogen could significantly reduce ventilation costs. Sandia National Labs has developed a metal hydride system for hydrogen storage at moderate pressures (Fig. 7.7).

Miscellaneous: Bicycles and scooters, especially prevalent in Asian cities, are prime candidates for FC technology to improve air quality and reduce noise

Fig. 7.5 Aircraft Prototype. *Source:* AA+W/Adobe Stock Photos

Fig. 7.6 Fuel Cell Fork Lift. *Source:* Toyota Material Handling

pollution. Companies like Honda and Yamaha have tested hydrogen FC scooters with on-board hydrogen storage. Wheelchairs are another area where FC technology can enhance mobility, with existing prototypes showcasing increased torque and acceleration, and a single hydrogen fill allowing for over 12 hours of operation.

Fig. 7.7 Hydrogen Fuel Cell Powered Train. *Source:* scharfsinn86/Adobe Stock Photos

7.2.6 Micropower Systems

Micropower systems, such as portable FCs, offer a significant advantage over traditional batteries by providing higher energy density, extended operational time, numerous recharge cycles, and reduced environmental impact during recycling. The prevalent technologies in micropower sources are direct methanol fuel cells (DMFCs) and PEMFCs. These technologies are increasingly being utilized as power sources for consumer electronics and communication devices, including digital cameras, mobile phones, laptops, radios, portable televisions, and night-vision goggles.

Japanese firms DoCoMo and Aquafairy have introduced water-based microfuel cell systems. The PEMFC for portable applications operates with a fuel cartridge, containing a hydrogen-producing catalyst, which supplies hydrogen gas to the anode of the PEMFC, while oxygen is drawn from the air for the cathode. The electricity generated by these cell power devices such as mobile phones or laptops. Figure 7.8 illustrates the chemical reactions within the FC and details the cartridge's function. These mobile power generators can deliver about 2 W of power and recharge devices multiple times with a recharge duration comparable to standard battery charging.

$$\text{Anode: } H_2 \rightarrow 2H^+ + 2e^-$$

$$\text{Cathode: } \tfrac{1}{2}O_2 + 2H^+ + 2e^- \rightarrow H_2O$$

$$\text{Whole Reaction: } H_2 + \tfrac{1}{2}O_2 \rightarrow H_2O$$

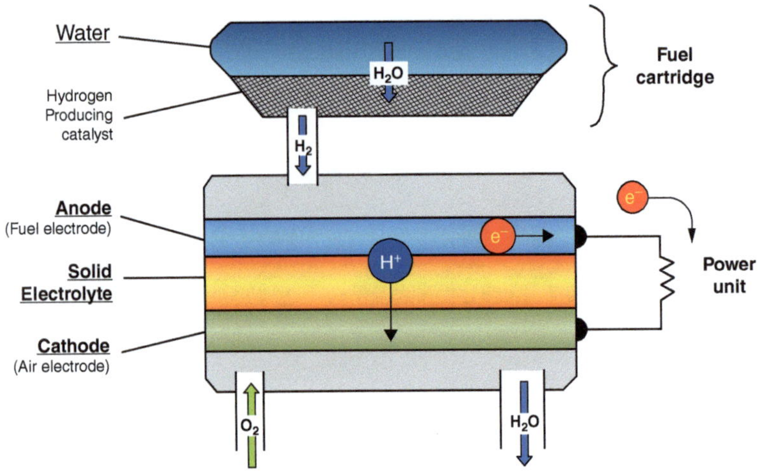

Fig. 7.8 The Chemical Reaction in Micropower PEMFC

Toshiba's compact DMFC, recognized by the Guinness Book of World Records in 2005 as the smallest of its kind, exemplifies the trend toward miniaturization in micropower FC supply. This diminutive DMFC, about the size of a woman's thumb, can power various electronic devices for approximately 20 hours on a single charge.

In the United States, Motorola has developed a 20 W DMFC for portable applications, integrating a fuel processor with a high-temperature FC and the necessary balance of plant (BoP) components. Another example is the Mobion technology by MTI Microfuel Cells, which features air-breathing operation, directs 100% methanol feed, and eliminates the need for water management, all within a compact 40 cc volume. Medis Technologies is nearing the final production stage of the "24/7" Power Pack, a portable FC designed to power a diverse range of electronic devices. The DMFC system diagram is presented in Fig. 7.9.

7.2.7 Mobile and Residential Power Systems

FC technology facilitates the creation of power generators of various sizes, particularly useful in regions prone to natural disasters such as typhoons, earthquakes, and floods, which often disrupt electrical power. FCSs also serve as reliable stationary backup systems where continuous power is essential, such as telecommunications, utilities, hospitals, industrial settings, and information technology sectors.

For residential use, the industry is developing small stationary systems of 10 kW or less, primarily based on SOFC or PEM technologies. The 1 kW units

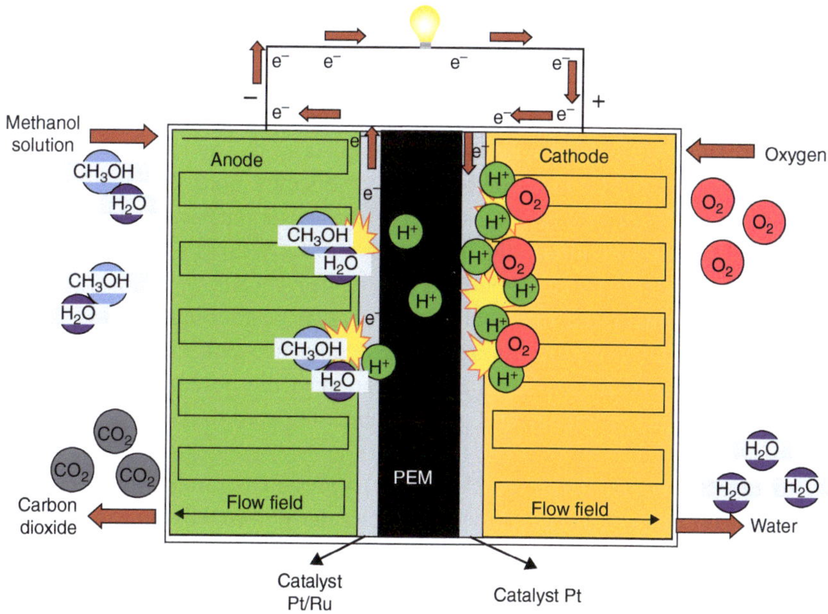

Fig. 7.9 Direct Methanol Fuel Cell (DMFC) Active System. *Source:* PEM Performance for Direct Methanol Fuel Cell Application: Perspective on Morphological Structure – Scientific Figure on ResearchGate, 2020/MDPI/CC BY 4.0

typically supply electricity for home appliances and hot water, while the larger 5 kW units are designed to provide continuous power to homes, with the grid covering only peak demand periods. These units are primarily fueled by natural gas, propane, or liquid petroleum gas (LPG), with a growing trend toward incorporating biomass.

In the United States, Plug Power's GenCore® fuel cell systems offer off-grid continuous power for residential use and as backup power sources, delivering up to 5 kW of electrical power across diverse operating conditions.

7.2.8 FCs for Space and Military Applications

Space Exploration: FCs have played a pivotal role in space missions, providing a strong impetus for their development. The Apollo missions utilized alkaline FC technology, and more recent space shuttles have employed PEMFCs for powering equipment. NASA is advancing FC technology to endure the rigorous conditions of space transport, aiming for energy and volume densities greater than 500 Wh/kg and 400 Wh/L to enable extended space missions. FCs are being considered for manned Mars missions, lunar bases, and unmanned high-altitude satellite aircraft, with the next-generation space shuttle likely to heavily incorporate FC technology.

Army: FC technology offers the military enhanced logistics, functionality, and capabilities, such as stealth operations and rapid start-up. The low thermal and acoustic signatures of FCs make detection by adversaries more challenging. The by-product water from FCs is also beneficial for various purposes. Experiments are underway to explore alternative power sources for weapons, electronic systems, and equipment to assist soldiers in carrying heavy loads (Fig. 7.10).

FCs are being integrated into night vision, global positioning systems, laser range finders, digital communication systems, intelligence-gathering sensors, and target designators. Hydrogen-based transportation in the military is also being evaluated for its fuel efficiency and acceleration compared to diesel or gasoline vehicles.

Aerospace: Unmanned aerial vehicles (UAVs) represent a promising application for FCs. AeroVironment has developed Helios, a solar-powered flying wing with a FCS for nocturnal flight, generating approximately 15 kW to power the aircraft's electric motors. Additionally, AeroVironment's "Hornet" is a FC-powered micro air vehicle equipped with a video camera for military surveillance.

Fig. 7.10 1.0 kW Metal Hydride Fuel Cell System. *Source:* Renewableandalternativeenergy/Wikimedia Commons/CC BY 4.0

Naval: Submarines powered by FCs offer high efficiency, low operating temperatures, and silent operation, making detection via sonar, noise, or infrared sensors challenging. FCs are also being considered for powering remote instruments and reconnaissance sensors that require very low power levels, as well as for various ships' onboard systems.

7.2.9 Maritime Applications

FCs are increasingly being recognized for their potential in maritime applications, particularly in commercial shipping and passenger ferries. The maritime industry, under pressure to reduce greenhouse gas emissions and improve air quality, is exploring FCs as a viable alternative to traditional diesel engines. In commercial shipping, FCs offer the advantage of significantly lower emissions while maintaining operational efficiency. They are particularly suitable for short-sea shipping routes where the regularity of operations allows for efficient hydrogen refueling infrastructure. Passenger ferries powered by FCs provide a cleaner, quieter, and more environmentally friendly travel experience, with several pilot projects already underway in various parts of the world, demonstrating the feasibility and benefits of this technology in maritime transport.

7.2.10 Wearable Technology and Internet of Things (IoT)

The integration of FC technology into wearable devices and IoT sensors represents a groundbreaking advancement in power solutions for these applications. Wearable devices, such as health monitors and smartwatches, benefit from the extended operational life and reduced charging frequency afforded by miniature FCs. This technology addresses the critical need for longer-lasting power sources in wearables, enabling continuous monitoring and data collection without the inconvenience of frequent recharging. Similarly, FCs are becoming an attractive option for powering sensors in the IoT ecosystem. These sensors, often deployed in remote or hard-to-reach locations, require reliable, long-lasting power sources. FCs provide a sustainable solution, reducing the need for battery replacements and ensuring consistent operation over extended periods.

7.2.11 Unmanned Underwater Vehicles (UUVs)

FCs are emerging as a key power source for UUVs, in both military and research applications. In military contexts, UUVs equipped with FCs benefit from extended range and operational duration, crucial for surveillance, reconnaissance, and mine countermeasure missions. The low acoustic signature of FCs also enhances the stealth capabilities of these vehicles. In the realm of oceanographic research, FCs enable UUVs to undertake prolonged missions, collecting valuable data on ocean currents, marine biology, and underwater geology. The ability of FCs to provide reliable power in the challenging underwater environment marks a significant advancement in the capabilities of UUVs.

7.2.12 Emergency and Disaster Relief

FCs have proven to be invaluable in emergency and disaster relief operations, offering portable and reliable power sources when traditional infrastructure is compromised. Portable FCs can be rapidly deployed to disaster zones, providing immediate power for critical communication equipment, medical facilities, and lighting. Their lightweight and compact design make them ideal for situations where mobility and quick setup are essential. In addition to their portability, FCs operate quietly and produce no harmful emissions, making them suitable for use in confined or densely populated areas. The resilience of FCs in varying environmental conditions further underscores their suitability for disaster relief efforts, ensuring a consistent power supply in the most challenging circumstances.

7.3 Electric Batteries

The current trend toward electrification in mobility relies heavily on energy storage devices, such as batteries for electric transportation and energy-producing FCs. This section focuses on the battery market, laying the groundwork for a subsequent discussion on the essentials of FCs.

7.3.1 History

Historical records suggest that as early as 200 BCE, primitive batteries were utilized in Iraq and Egypt for electroplating and gilding precious metals. The era of modern batteries commenced around 1800 with Alessandro Volta's groundbreaking discovery of electrical current generation through chemical reactions between dissimilar metals. This pivotal moment spurred decades of experimentation, significantly advancing the field of electrochemistry.

By the 1860s, notable advancements were made by Georges Leclanché with the carbon–zinc wet cell and Raymond Gaston Planté with the lead-acid battery, despite their limitations in shelf life. The landscape of battery technology further evolved in 1881 when Émile Alphonse Faure introduced an improved battery design using a mixture of lead oxides, enhancing reaction speed and efficiency.

The turn of the century saw Thomas Edison's development of the nickel storage battery and, shortly thereafter, the nickel–iron battery, marking significant milestones in battery innovation.

The mid-20th century introduced the small alkaline battery, followed by the development of the improved alkaline–manganese battery. The discovery of the solar cell in 1954 and the hydrogen–oxygen FC in 1956 broadened the applications of batteries. The 1960s invention of the gel-type electrolyte lead-acid battery and the commercialization of lithium-ion batteries in the 1990s revolutionized portable electronics, telecommunications, and space exploration. Innovations such as temperature monitoring sensors and power flow control chips have since enhanced battery life and efficiency. Currently, research into incorporating carbon nanotubes promises to further extend the lifespan of traditional batteries, signifying ongoing progress in battery technology.

The advent of the automobile introduced simple mechanisms for operation, including bells for signals, gas-powered headlights, and mechanical engine cranking. The 1920 invention of the electric starter marked a pivotal shift, significantly increasing the reliance on batteries for transportation. The development of the sealed battery in 1971 simplified maintenance, enhanced safety, and extended the lifecycle of batteries, representing a significant advancement in battery technology.

Initially, all vehicles operated on a 6 V electrical system until the mid-1950s, when the growing demands for ICE performance prompted a transition to 12–13.6 V systems. These batteries, crucial for starting, lighting, and ignition (SLI) in ICE vehicles, underwent attempts to further increase voltage to 42 V to accommodate more efficient engines, complex electronic ignitions, and the burgeoning array of sensors and actuators. However, this transition has been stalled by the durability challenges of electrical system components.

The evolution of battery technology continues to play a critical role in the development of electric propulsion systems, with increasing efforts to enhance battery voltage and efficiency to meet the demands of electric vehicles (EVs). In parallel, the utility sector explores a diverse array of battery technologies for large-scale energy storage solutions. Initially dominated by lead-acid batteries, the landscape now includes sodium sulfide (NaS) and lithium-ion batteries, which offer improved performance and are becoming increasingly commercially viable.

Batteries, as electrochemical cells, facilitate energy storage and release through reversible chemical reactions. This capability is essential for both discharge and recharge cycles, though challenges such as cost, environmental concerns related to disposal, and the finite number of recharge cycles persist. Addressing these issues is crucial for maximizing the value and sustainability of battery technologies in both transportation and utility-scale energy storage applications (Figs. 7.11 and 7.12).

Fig. 7.11 Rechargeable Battery Banks. *Source:* Jelson25/Wikimedia Commons/CC BY 3.0

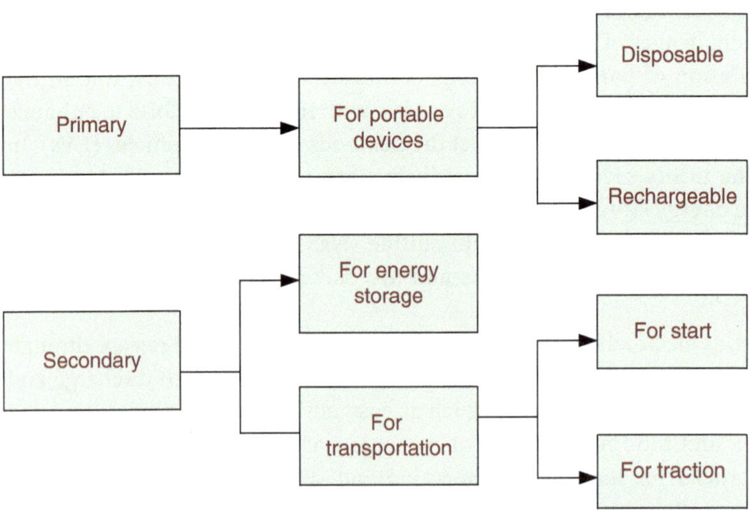

Fig. 7.12 Batteries Classification

7.3.2 Primary Batteries

Primary batteries, also known as alkaline dry cells, are integral to a multitude of consumer applications characterized by low current drain and relatively brief usage durations. These batteries consist of an electrolyte, usually in gel form, positive and negative electrodes, and external terminals connected to the electric load. The electrolyte's role as an ionic conductor facilitates electron transfer between electrodes. The voltage or electromotive force generated by these batteries is determined by the chemical properties of the materials used, independent of the electrodes' size or the quantity of electrolyte. Structurally, a battery is designed as a dielectric container housing internal electrolytic cells and electrodes linked to terminals.

Batteries are broadly categorized into disposable and rechargeable types. Primary cells belong to the disposable category, designed to be used until the chemical reactions producing electrical current cease, at which point they must be discarded. Despite this, there are rechargeable options within the primary battery family, commonly utilized in portable devices such as telecommunications equipment, laptop computers, and medical instruments. These are favored for their convenience in situations without access to alternative power sources.

Further classification distinguishes batteries as either dry cell or wet cell. Dry cells confine the electrolyte within a porous material to prevent it from flowing, while wet cells allow the liquid electrolyte to move freely within the enclosure. Carbon–zinc dry cells, for example, are favored for powering small devices like pocket radios, CD players, flashlights, and mobile phones due to their compactness and efficiency. Conversely, alkaline batteries, which employ a hydrous alkaline solution as the electrolyte, offer cost-effective production and enhanced reliability. Their reduced corrosiveness compared to zinc-based alternatives extends their lifespan, making them a preferred choice for a wider range of applications.

Disposable primary cells are capable of producing current immediately upon assembly. However, they are not designed for reliable recharging due to the irreversible nature of their chemical reactions. After discharge, the active materials within the battery may not return to their original state, limiting the cell to a single-use lifecycle. This inherent limitation underscores the critical need for sustainable disposal practices and ongoing research into more environmentally friendly alternatives that can match the convenience and utility of primary batteries without the accompanying waste.

7.3.3 Batteries for Portable Devices

The proliferation and miniaturization of portable consumer electronics have necessitated significant advancements in battery technology, with lithium-ion (Li-ion) batteries leading the charge. Their widespread adoption stems from an

Fig. 7.13 A Battery Pack Charger for Android and iPhone. *Source:* JETE Official/Wikimedia Commons/CC BY-SA 4.0

exceptional power-to-weight ratio, high energy efficiency, robust high-temperature performance, minimal self-discharge rates, and full rechargeability. Li-ion batteries comprise multiple lithium-ion cells housed within a single enclosure, designed to deliver consistent power to portable devices. Integral to all rechargeable batteries are peripheral devices facilitating the reversed chemical reaction necessary for recharging.

Efforts in Li-ion battery development are increasingly focused on reducing costs, extending the number of charging cycles, and enhancing power density. These batteries dominate the consumer rechargeable market due to their reliability and performance advantages.

Beyond Li-ion, portable devices utilize various sealed dry cell types. Nickel–cadmium (NiCd) batteries, known for their use in power tools, two-way radios, and medical equipment, offer durability and reliability. Nickel–zinc (NiZn) and nickel–metal hydride (NiMH) alternatives provide higher capacities compared to NiCd, catering to a broader range of applications.

Recent innovations include smart battery packs equipped with embedded electronics, optimizing charging processes and ensuring battery protection. These advancements underscore the continuous evolution of battery technology, aiming to meet the growing demands for efficiency, sustainability, and performance in portable consumer electronics (Fig. 7.13).

7.3.4 Secondary Batteries

Secondary batteries stand out for their ability to undergo multiple recharge cycles after depletion. This characteristic is facilitated through the application of an external electrical current, reversing the chemical reactions that occur during discharge. Essential to this process are external chargers that bridge the power source – usually the electric grid – with the battery, adapting the high grid voltage to the lower voltage needed for safe battery recharging.

Conceptually, all automotive batteries fall within the rechargeable category, with their applications bifurcated into three primary functions: energy storage, engine starting, and vehicle traction. Energy storage batteries, which may contain liquid or gel electrolytes, are composed of cells connected in series. Each cell houses alternating positive and negative plates, immersed in an electrolyte that acts as a conduit for ion exchange. During the recharging phase, the operation reverses from its discharging mode, effectively converting electrical energy back into chemical energy for storage. These batteries are crucial in providing energy storage solutions for alternative power generation, such as solar panels and wind turbines, enabling a consistent power supply even in the absence of direct sunlight or wind.

The secondary starting battery, ubiquitous in transportation, delivers the necessary electric power for engine ignition, lighting, and the ignition system, collectively known as SLI batteries. Post-engine start, the alternator, generating between 13.5 and 14.5 V, takes over the vehicle's power supply. SLI batteries are engineered to emit a substantial current surge briefly, subsequently recharged by the vehicle's onboard alternator. While lead-acid batteries have traditionally dominated this space, the shift toward lithium-ion batteries for starting applications reflects ongoing innovation, offering weight savings and improved efficiency.

Battery technology is characterized by several critical parameters that define its performance in various applications:

- *Specific Energy (Energy Density)*: This metric refers to the amount of energy stored in a battery relative to its mass, expressed in Watt-hours per kilogram (Wh/kg). High specific energy is crucial for applications where lightweight is essential, such as in EVs and portable electronic devices.
- *Power Density*: Also known as volume-specific power, this parameter measures the power output per unit volume of the battery, denoted as Watts per cubic meter (W/m^3). Power density is vital for applications requiring batteries to deliver high power bursts, such as in starting engines or during peak demand periods in grid storage.
- *Battery Storage Capacity*: This measurement allows for the comparison of the charge-holding capabilities of different batteries, quantified in Ampere-hours (A-h) and Cranking Amperes (CA). CA indicates the current a battery can deliver at 32 °F (0 °C), while Cold Cranking Amperes (CCA) measures the current output at 0 °F (−18 °C), crucial for evaluating a battery's performance in cold conditions.
- *C-rates*: The storage capacity is further detailed by C-rates, defining the relationship between discharge/charge current and the battery's nominal capacity. For instance, at a 1C rate, a battery with a 10 A-h rating should deliver 10 A for 1 hour, whereas at 0.5C, it would provide 5 A for 2 hours.

Table 7.1 Specific Energy and Energy Density Data

Storage Material	Specific Energy (MJ/kg)	Energy Density (MJ/L)	Direct Uses
Alkaline battery	0.5	1.3	Portable electronic devices, flashlights
Lithium battery (nonrechargeable)	1.8	4.32	Portable electronic devices, flashlights
Lead-acid battery	0.17	0.56	Transportation
Lithium-ion battery	0.36–0.875	0.9–2.63	Laptop computers, mobile devices, electric vehicles
Nickel–metal hydride battery	0.288	0.504–1.08	Laptop computers, mobile devices, electric vehicles

Source: Adapted from Energy Density, Wikipedia, https://en.wikipedia.org/wiki/Energy_density, last accessed 26 March 2024

Table 7.1 presents comparative data on Specific Energy and Energy Density for various storage technologies, offering insight into the advancements and efficiency of different battery types.

7.3.5 Traction Battery

The traction battery, or electric vehicle battery (EVB), plays a pivotal role in powering electric transportation devices. These batteries are integral to the operation of a wide array of EVs, including forklifts, golf carts, motorcycles, and both passenger and commercial EVs. Initially, EV development heavily relied on automotive lead-acid rechargeable batteries due to their availability and cost-effectiveness. However, advancements in battery technology have shifted the preference toward lithium-ion and NiMH batteries, mirroring the trends observed in consumer electronics.

Lithium-ion batteries have become the predominant choice for new plug-in hybrid electric vehicles (PHEVs) and EVs, offering superior performance over the NiMH batteries traditionally used in hybrid electric vehicles (HEVs). The distinction between EVBs and SLI batteries lies in their operational purpose. While SLI batteries are designed to deliver short bursts of high power for starting engines, traction batteries must sustain power delivery over extended periods, necessitating high capacity (measured in ampere-hours or coulombs), high power-to-weight ratio, and high energy density.

The driving range of EVs is directly influenced by the specific energy of the EVB. The competitive edge of EVs against ICE vehicles hinges on the continuous development of battery technologies. Achieving specific energy levels comparable to those of liquid petroleum-based fuels is crucial for the broader adoption and success of EVs in the transportation sector.

7.3.6 Rechargeable Batteries

Rechargeable batteries, or "secondary cells," are essential components in a wide range of applications, from small-scale consumer electronics to large megawatt grid systems. Characterized by their ability to undergo reversible electrochemical reactions, these batteries offer a sustainable alternative to disposable ones, presenting lower total costs of use and reduced environmental impacts.

Despite their higher upfront costs, rechargeable batteries can be recharged at a minimal expense and utilized multiple times, making them a cost-effective solution in the long run. They are available in various form factors, including those comparable to disposable batteries, allowing for easy integration into existing devices.

Common Rechargeable Battery Chemistries:

- *Lead–Acid Battery*: Dominating the electric storage market, a single lead-acid cell generates about 2 V when charged. It comprises a metallic lead negative electrode and a lead sulfate positive electrode, immersed in dilute sulfuric acid. While offering low maintenance and cost, lead-acid batteries face limitations in energy density and lifespan due to rapid discharge rates.
- *Nickel–Cadmium Battery (NiCd)*: Known for using nickel oxide hydroxide and metallic cadmium as electrodes, NiCd batteries have been largely phased out in favor of more environmentally friendly alternatives due to cadmium's toxicity and subsequent regulations.
- *NiMH Battery*: Introduced commercially in 1989, NiMH batteries have become prevalent in consumer and industrial applications. They feature a hydrogen-absorbing alloy as the negative electrode, offering a more eco-friendly profile compared to NiCd batteries.
- *Lithium-Ion Battery*: Favored for their excellent energy-to-mass ratios and low self-discharge rates, lithium-ion batteries are the primary choice in consumer electronics, providing high efficiency and long cycle life.
- *Aluminum-Sulfur Battery*: An emerging technology utilizing earth-abundant materials like aluminum and sulfur with rock salt crystals as the electrolyte. This innovation promises a cheaper, more sustainable alternative to traditional lithium-based batteries, potentially revolutionizing energy storage solutions.

Rechargeable batteries' evolution reflects ongoing efforts to enhance energy storage technologies, offering diverse options tailored to specific needs and applications. As research continues, the development of more sustainable, efficient, and cost-effective battery chemistries remains a pivotal focus in meeting global energy demands.

7.3.7 Lead-Acid Batteries

Lead-acid batteries, a staple in the realm of energy storage, are renowned for their high power output, cost-effectiveness, safety, and reliability. Primarily installed in ICE transportation units, these batteries offer substantial advantages, albeit with considerations regarding their operational depth of discharge. Full discharges are notably detrimental, potentially reducing the battery's overall lifespan.

In recent years, gel-type lead-acid batteries have begun to supplant traditional wet-cell versions. This transition is driven by the gel-type's semisolid electrolyte, which not only simplifies maintenance but also enhances durability. The advent of fully sealed dry-cell devices has expanded the utility of portable rechargeable lead-acid batteries, making them indispensable in mobile phones, laptops, and a range of medical and industrial instruments.

During the initial exploration of EV technology, vehicles like the EV1 prototype predominantly utilized lead-acid batteries as traction batteries. This choice was predicated on the technology's maturity, widespread availability, and affordability. Lead-acid batteries bifurcate into two primary categories: starter batteries for automobile engines and deep-cycle batteries. Starter batteries are engineered to accommodate high charge rates from alternators for swift recharging. Conversely, deep-cycle batteries cater to the sustained energy demands of EVs, such as fork-lifts and golf carts, and serve as auxiliary power sources in airport transportation and recreational vehicles.

However, the utilization of lead-acid batteries in EVs introduces significant challenges, notably their substantial weight, which can constitute up to 40% of a vehicle's total mass. Compared to petroleum fuels, lead-acid batteries exhibit markedly lower energy densities, approximately 30–40 Wh/kg, necessitating larger battery packs to achieve comparable ranges. Additionally, the efficiency and storage capacity of lead-acid batteries diminishes at lower temperatures, with the necessity to power heating coils further reducing efficiency and operational range by as much as 40%.

Lead-acid batteries, a foundational technology in energy storage, have long been utilized in ICE transportation units for their high power output, cost-effectiveness, safety, and reliability. Traditional lead-acid batteries emit trace amounts of hydrogen, oxygen, and sulfur during charging and operation, which are typically harmless with proper ventilation. Modern advancements have led to the production of sealed lead-acid batteries, effectively eliminating the issue of gas emission.

The anatomy of a lead-acid battery includes alternating plates of lead (cathode) and lead dioxide (anode) submerged in a sulfuric acid solution electrolyte. In a standard 12 V battery, 6 cells each produce approximately 2 V. The electrochemical reaction that occurs within these cells releases electrons, facilitating the flow of electricity. Discharging the battery leads to the conversion of the plate materials into lead sulfate due to the reaction with the electrolyte acid. Recharging reverses this process, restoring the plates to their original state and allowing the cycle to be repeated numerous times.

Lead-acid batteries are renowned for their low production cost and capability to deliver high surge currents, making them the preferred choice as SLI batteries in modern automobiles. These batteries are designed to provide a peak current sufficient for engine SLI systems, with automobile alternators supplying high charge rates for rapid recharging.

Despite their widespread use, the application of lead-acid batteries in EVs is constrained by several factors: sensitivity to cold temperatures, packaging limitations, excessive weight, and a relatively short cycle life. However, recent advancements in battery efficiency, capacity, material innovation, safety, and durability have begun to address these challenges, enhancing their potential for EV applications.

Key areas for improvement in lead-acid battery technology for EV use include

- *Driving Range*: Limited capacity restricts EV driving range. Typical lead-acid battery packs, storing 12–15 kW-hours, offer ranges of only about 50 miles. Efforts to increase capacity and optimize efficiency are crucial for extending the practical range of EVs.
- *Recharge Time*: The lengthy recharge duration of lead-acid batteries, requiring several hours for a full charge, poses challenges for EV usability. Innovations aimed at reducing recharge times while maintaining battery integrity are essential for user convenience.
- *Cost and Lifespan*: The significant expense associated with large battery packs and their need for replacement during a vehicle's operational life adds to the total cost of ownership. Enhancing the durability and extending the cycle life of lead-acid batteries can mitigate these financial burdens.
- *Bulk and Weight*: The substantial weight and volume of lead-acid battery packs affect vehicle design and efficiency. Research into lighter and more compact battery solutions is ongoing, aiming to improve vehicle performance and energy efficiency.

Ongoing research and development are geared toward overcoming these limitations, with the goal of making lead-acid batteries a more viable option for the evolving EV market. Their historical significance and adaptability underscore the potential for lead-acid batteries to contribute significantly to the future of sustainable transportation.

7.3.8 Lithium-ion Batteries

Lithium-ion (Li-ion) batteries have emerged as a cornerstone of energy storage technology across various domains, notably in portable devices, EVs, and even stationary applications. Their ascendancy is attributed to their high energy density, efficiency, and longevity. Despite the geographical limitation of lithium resources, intense development efforts are underway, promising imminent advancements in cost reduction, enhanced lifespan, and safety improvements.

Contrasting with the well-established lead-acid batteries – predominantly utilized in automotive and stationary grid applications – lithium-ion batteries offer a cleaner alternative, mitigating the environmental and health concerns associated with lead toxicity. However, the competitive edge of lead-acid batteries lies in their lower investment costs and the extensive manufacturing ecosystem.

Lithium-ion technology, particularly for EVs, has made significant strides, utilizing lithium cobalt oxide cathodes and graphite anodes. This configuration maximizes the storage of lithium ions per unit surface area, surpassing the capacity of other battery technologies. Key attributes of lithium-ion batteries include an energy density exceeding 200 Wh/kg, superior power density, and up to 90% charge/discharge efficiency. Nonetheless, challenges persist, including limited charging cycles, lifespan degradation, and safety risks associated with fire if the batteries are damaged or charged improperly.

Heat generation stands as the primary concern for fire safety and battery failure, often exacerbated by accelerated corrosion within the battery. Corrosion at the battery terminals can increase electrical resistance, with sulfate accumulation occurring when batteries remain undercharged for extended periods. This phenomenon underscores the need for rigorous maintenance and monitoring to prevent efficiency loss and potential safety hazards.

As lithium-ion battery technology continues to evolve, addressing these limitations remains paramount. Ongoing research focuses on improving charge cycle capabilities, mitigating degradation effects, and enhancing thermal management to reduce fire risks. Advances in materials science and battery design are crucial for overcoming the current barriers, paving the way for safer, more reliable, and cost-effective lithium-ion batteries. These improvements are vital for expanding the role of lithium-ion batteries in sustainable transportation, renewable energy storage, and beyond, marking a significant step toward a greener future.

The advancement of lithium-ion battery chemistry is a balancing act, aiming to optimize between critical factors such as safety, energy and power density, environmental sustainability, rapid charging capabilities, and the overall lifespan of the battery. Innovations in this field are making remarkable strides, as exemplified by the A123 battery system, renowned for its use in portable tool charging. This particular battery boasts over 1,000 charge cycles, exceptional charge and discharge rates, and the anticipation of a 10-year lifespan, supporting up to 7,000 charge cycles with its lithium iron phosphate composition.

In the quest for higher energy densities, lithium vanadium oxide has shown promising results, especially in automotive applications like the Subaru prototype

Fig. 7.14 2012 Chevrolet Volt T-shaped Lithium-Ion Battery. *Source:* Data Centre Dynamics Ltd, https://media.datacenterdynamics.com/media/images/2012-Chevrolet-Volt-battery-pack.original.jpg, last accessed 01 March 2024

G4e ("green for the earth"), which has seen its energy density doubled, thereby extending the driving range by up to 40%. Efforts to enhance battery performance further involve innovative replacements for the traditional graphite anode. The employment of nanowires, for instance, substantially increases electrode surface areas, while the integration of silicon nanoparticles as anodes and the development of composite materials for cathodes are paving the way for significant improvements in energy density.

These advancements reflect the dynamic nature of lithium-ion battery research, focusing on enhancing efficiency, safety, and environmental compatibility. As these technologies evolve, they promise to revolutionize energy storage solutions across various sectors, from consumer electronics to EVs, marking a significant leap toward more sustainable and efficient power sources (Fig. 7.14).

7.3.9 NiMH and High-Temperature Batteries

NiMH batteries, alongside high-temperature battery technologies such as sodium–nickel–chloride ($NaNiCl_2$, also known as zebra batteries) and sodium-sulfur (NaS), represent significant advancements in energy storage solutions. High-temperature batteries, operating between 270 and 350 °C, require meticulous thermal management to maintain optimal temperatures during charge and discharge cycles, necessitating sophisticated insulation designs. These batteries find their niche in stationary applications like peak shaving and load shifting, offering an alternative to conventional lead-acid and lithium-ion batteries. However, their global distribution remains limited, primarily due to the relatively few manufacturers and significant safety concerns, particularly the risk of fire incidents associated with NaS batteries.

NiMH batteries, renowned for their use in computers and medical equipment, provide a balanced specific energy and power output. They surpass lead-acid batteries in lifecycle duration and are recognized for their safety and tolerance to abuse. NiMH batteries have proven their efficacy in both EVs and HEVs, showcasing their versatility. Despite their benefits, challenges such as high initial costs, notable self-discharge rates, and heat generation at elevated temperatures persist. Moreover, the management of hydrogen loss remains a crucial aspect of their operation.

Established as a reliable technology, NiMH batteries exhibit an energy density between 40 and 80 Wh/kg – superior to that of lead-acid batteries but with a lower efficiency range of 60–70% in charging and discharging. Their longevity is evidenced in applications such as the Toyota RAV4 Hybrid, where NiMH batteries continue to perform after 100,000 miles and over a decade of usage. However, drawbacks include lower efficiency compared to other battery types, complex charging requirements, and diminished performance in cold conditions, posing challenges for broader adoption.

7.3.10 Molten Salt Battery

Molten salt batteries, particularly the sodium or "zebra" battery, utilize molten chloroaluminate sodium ($NaAlCl_4$) as their electrolyte, a composition sometimes dubbed "hot salt." As a relatively mature technology, the Zebra battery distinguishes itself with an energy density of 120 Wh/kg and possesses a reasonable series resistance, making it a viable option for EVs. Capable of enduring a few thousand charge cycles, Zebra batteries are notable for their nontoxic nature, aligning with environmental sustainability goals.

However, the Zebra battery faces challenges in terms of power density, which is less than 300 W/kg, and operational requirements, as the electrolyte must be maintained at approximately 270 °C (520 °F) to remain functional. This necessity for constant heating not only leads to energy wastage but also complicates the long-term storage of charge, posing logistical and efficiency issues. Despite these drawbacks, the high energy density and environmental friendliness of Zebra batteries make them a subject of ongoing interest for applications in EVs and beyond, as advancements continue to address their limitations.

7.3.11 Recharging

The advent of EVs has necessitated the development of efficient and accessible recharging solutions. Battery electric vehicles (BEVs) predominantly recharge from the power grid, which utilizes a mix of renewable and nonrenewable energy sources including coal, hydroelectricity, and nuclear power. An alternative to traditional grid charging is the establishment of public and private charging stations designed for general use or specific fleet segments, such as taxi services.

7.3.11.1 Charging Infrastructure

Personal charging sources, installed at homes or public locations and powered by renewable sources like photovoltaic solar cells, micro-hydro, or wind turbines, are gaining popularity due to global warming concerns. With appropriate power supplies, optimal battery lifespan is usually maintained at charging rates not exceeding "0.5C," resulting in a full charge within two to three hours. However, advancements in technology have enabled faster charging capabilities.

The charging time is constrained by the supply connection's capacity and the battery's maximum charge rate. Standard household outlets deliver 1.5 kW in countries with 110 V supply systems and 3 kW in regions with 240 V supply. Even with access to higher power supplies, the charging process is limited by the battery's permissible C-rate, allowing for recharge times of 20–30 minutes to 80% capacity for some batteries.

For consumer products, minimizing charge time is crucial, with research efforts aimed at reducing it to as little as 60 seconds. In transportation, while quick recharging is desirable, it is not critical as charging can occur overnight or during work hours. The trend toward constructing special parking bays for EV charging in metropolitan areas is growing. Companies like Tesla are developing home and office battery charging equipment and establishing a network of high-powered Superchargers across North America, Europe, and Asia.

Charging power can be connected to the car via two methods:

1) *Conductive Coupling*: This method involves a direct electrical connection through special high-capacity cables equipped with weatherproof sockets and protection for drivers from high voltage. Standards such as the SAE 1772 conductive connector have been developed for plug-in vehicles.
2) *Inductive Charging*: This approach utilizes a special connector inserted into a slot on the car, functioning as one winding of a transformer, with the other built into the vehicle. When the connector is inserted, it completes a magnetic circuit that powers the battery pack. Inductive charging systems, where one winding is attached to the underside of the car and the other remains on the garage floor, offer the advantage of eliminating electrocution risks due to the absence of exposed conductors. Safety features such as interlocks, special connectors, and ground fault detectors can make conductive coupling nearly as safe.

7.3.11.2 EV Range and Battery Evolution

The General Motors EV1, equipped with NiMH batteries in 1999, had a range of 75–150 miles. Contemporary BEVs boast ranges up to 350 miles, depending on the number and types of batteries used, signifying significant advancements in battery technology and EV efficiency.

As EV adoption continues to rise, the evolution of recharging infrastructure and technology will play a crucial role in supporting the transition to sustainable transportation, balancing efficiency, safety, and environmental considerations.

7.3.12 Battery Specifications

The rapid advancement in battery technology complicates the comparison of different batteries due to the diversity in specifications and performance metrics. For instance, the Chevy PHEV Volt exemplifies modern efficiency with its 120-V portable cord that achieves a full recharge in approximately 13 hours, and a more rapid recharge time of about 4.5 hours with a 240-V charging system. The 2017 V model boasts an impressive driving range, capable of traveling up to 53 miles on a fully charged Li-ion battery without gasoline and extending to 420 miles with a combination of a full charge and a full tank of gas.

Nissan Leaf's traction battery (Table 7.2) further illustrates the detailed specifications required for rechargeable EVs, highlighting the complexities in assessing and comparing battery performance across different vehicle models.

7.3.12.1 Environmental Impact and Recycling

Batteries exert a significant environmental impact throughout their lifecycle, encompassing production, usage, disposal, and recycling phases. Given the potential hazards associated with used batteries, they are classified as hazardous waste. Remarkably, the 12-V car lead-acid battery stands as the world's most recycled product, with recycling rates exceeding 95% in the United States and nearly 97% of the lead being recoverable, showcasing a successful model of sustainability.

As EVs gain popularity, the demand for battery recycling infrastructure is expected to grow, mitigating the risk of hazardous materials entering the waste stream. Current recycling methods include:

- *Smelting*: An extractive metallurgical process operating at high temperatures; smelting allows for the recovery of valuable metals while utilizing electrolytes and other organic components as fuel. This process is widely implemented on a large scale for various battery types, including lithium-ion and NiMH.
- *Direct Recovery*: A more energy-efficient, low-temperature process that separates components through physical and chemical methods, allowing for the

Table 7.2 Example of Rechargeable Traction Battery Specs Used in Nissan Leaf

Type	Laminated Lithium-Ion Battery
Voltage	403.2 V
Nominal voltage	360 V
Total capacity	24 kWh [30] (16 kWh available, 67% DoD [31], 21 kWh declared [32])
Power output	Over 90 kW
Energy density	140 Wh/kg [33]
Power density	2.5 kW/kg [33]
Dimensions	61.8 × 46.8 × 10.4 in. (1570.5 × 1,188 × 264.9 mm) [29]
Weight	648 lbs. [34]
Number of modules	48, each with four cells (a total of 192 cells) [30, 37]
Battery pack contents	Positive electrodes: lithium manganate Negative electrodes: carbon Cells Modules Assembly parts
Charging times	Quick charger DC50 kW (0–80%): approx. 30 min Home-use AC240 V charging dock (0–100%): 8 hours Regular 110/120 V 15-A outlet: 22 hours [39]
Battery layout	Under seat and floor

recovery of all active materials and metals. This approach promises a more sustainable alternative to traditional smelting, reducing energy consumption, and facilitating comprehensive material recovery.

The evolving landscape of battery technology, coupled with increased awareness and innovation in recycling methods, underscores the critical role of environmental stewardship in the sustainable advancement of battery-powered solutions.

7.4 Hydrogen Transportation

7.4.1 FC Reformers

In the context of FC technology, the issue of hydrogen sourcing and integration into FC structures is paramount. While previous sections have outlined various FCs utilizing hydrogen, it is crucial to delve into hydrogen production pathways and their practicality in fuel system coupling. Hydrogen, being a gaseous substance, presents significant challenges in terms of volume occupation and storage when considered for FCSs. The conventional method of using gas cylinders for hydrogen storage, though straightforward, proves impractical due to the sheer volume required for continuous electric power supply, necessitating frequent replenishment.

As previously introduced in section 4.2.1.8 in Chapter 4, to mitigate storage and distribution challenges, reformers emerge as a viable solution, converting readily available fuels like natural gas or alcohol into hydrogen through a chemical process. This process not only facilitates the decomposition of hydrocarbons or alcohols into hydrogen but also generates heat and other by-products that must be efficiently removed to prevent contamination of the hydrogen gas produced.

Comparatively, liquid fuels such as gasoline offer a higher energy density per unit volume than gaseous hydrogen. However, liquefying hydrogen to enhance its energy density involves maintaining extremely low temperatures, presenting significant logistical challenges for transportation. In contrast, reformers, particularly steam reformers, circumvent these issues by producing pure hydrogen from methanol or natural gas without the need for low-temperature storage.

The breakdown mechanism of methanol or natural gas within reformers is a critical aspect of this process, facilitating the efficient production of hydrogen for FCs. By leveraging reformers, the integration of hydrogen into FCSs becomes more feasible, addressing the pivotal concerns of storage and distribution while capitalizing on the higher energy density of liquid fuels. This technology plays a crucial role in advancing FC applications, offering a practical approach to harnessing hydrogen as a clean energy source for various applications.

7.4.2 Steam Reformer

The steam reforming process is a critical method for producing pure hydrogen, utilizing methanol and natural gas as primary feedstocks. When methanol, combined with water, is introduced into a heated chamber containing a catalyst, it undergoes decomposition to yield hydrogen and carbon monoxide (CO) through the chemical reaction:

$$CH_3OH(l) \rightarrow CO(g) + 2H_2(g)$$

The presence of water vapor in the reaction chamber facilitates a secondary reaction where CO interacts with water to produce hydrogen and carbon dioxide (CO_2), effectively converting the more toxic CO into the less harmful CO_2:

$$CO(g) + H_2O(g) \rightarrow CO_2 + H_2$$

Similarly, natural gas or methane can be converted into hydrogen through a parallel steam reforming process. Methane reacts with water vapor, generating CO and additional hydrogen:

$$CH_4\,(g) + H_2O\,(g) \rightarrow CO\,(g) + 3H_2\,(g)$$

Following this reaction, CO once again reacts with water vapor, yielding CO_2 and more hydrogen:

$$CO\,(g) + H_2O\,(g) \rightarrow CO_2\,(g) + H_2\,(g)$$

The hydrogen gas produced via steam reforming, however, contains trace amounts of CO and CO_2, which must be removed to prevent FC contamination. This purification step is essential for ensuring the hydrogen supplied to the FC is of high purity, optimizing the efficiency and longevity of FC operations.

7.4.3 Fuel Processor

Central to FCS architecture is the fuel processor, a unit designed to adapt various fuel types, including gas, liquid, or solid, for FC compatibility. The process begins with the crucial removal of impurities such as sulfur, halides, and ammonia, which can compromise electrode efficiency. Following this, the processor converts the purified fuel into hydrogen while also addressing and eliminating by-products like CO and CO_2 through a water–gas shift reaction. This dual-phase process underscores the fuel processor's role as an indispensable, cost-effective component of the FCS, with operational temperatures delineated in Fig. 7.15 to illustrate the precise requirements for effective fuel processing.

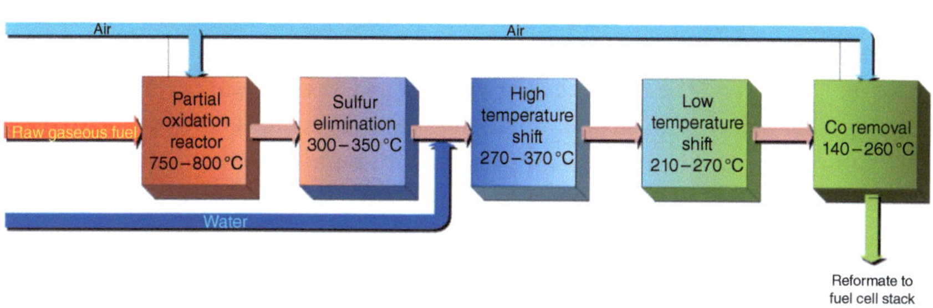

Fig. 7.15 Fuel Processor Temperature Requirements

7.4.4 FCS Architecture

In the realm of sustainable energy solutions, FC technology emerges as a cornerstone for various applications, from portable electronics to residential power and beyond. Having previously introduced FCS Architecture in section 4.3 in Chapter 4, the architecture of a fuel cell power system (FCPS) is meticulously designed to meet the demands of these applications, offering a sophisticated blend of subsystems working in harmony to generate electrical energy efficiently.

At its core, the FCPS aims to precisely control and supply reactants – namely water and gases – to facilitate the generation of electrical energy, while simultaneously managing excess heat and by-products. This is achieved through an intricate configuration of subsystems, each tailored to specific industrial uses. Currently, FCPS finds its application in four primary domains:

1) *Micropower Sources*: Essential for powering computers and communication devices, these systems are generally limited to outputs not exceeding 100 W.
2) *Residential or APUs*: Capable of delivering power up to 10 kW, APUs serve the energy needs of modern homes.
3) *Mobile Applications*: Designed for vehicular use, these FCs are typically restricted to outputs around 100 kW.
4) *Stationary Power Generation*: Encompassing a broad spectrum, these systems can scale to meet demands in the megawatt range, catering to large-scale power needs.

In micropower systems, fuel is stored either as a liquid or as gas within the fuel storage section, designed for easy replacement upon depletion. A critical component, the power conditioning unit, acts as an intermediary between the fuel storage and the FC, ensuring smooth operation during power fluctuations or system start-up. This configuration may also incorporate fuel delivery systems and energy storage devices, such as ultracapacitors, to stabilize power consumption.

For broader applications, the FCPS architecture integrates a fuel processor, FC stack, and various subsystems, all regulated by a balance-of-plant section and power conditioner. Removing excess heat and moisture is vital for optimal performance, with water recycling for FC stack humidification enhancing efficiency (Fig. 7.16).

The system typically divides into two parts: one housing the fuel processor and the other comprising the FC stack alongside the power conditioner. The processor generates hydrogen from diverse fuels like natural gas, gasoline, or methanol, with subsequent purification to meet the purity requirements of the FC stack. This stack, sensitive to impurities, relies on the power conditioning unit to convert the generated direct current (DC) voltage to alternating current (AC), making the electricity suitable for a wide array of applications.

PEMFC, favored for automotive applications, demonstrates an efficiency of 45% at low operating temperatures, boasting high power density. Residential systems often utilize natural-gas-based FCs, aligning with environmental standards and efficiency criteria.

The development of FCPS marks a pivotal advancement in FC technology, embodying the principles of sustainability, efficiency, and adaptability. From powering small electronic devices to providing residential energy solutions and propelling vehicles, the tailored subsystem configurations within FCPS ensure that FCs can meet diverse energy demands effectively. This modular approach underscores not only the versatility of FC technology but also its potential to significantly impact energy consumption patterns and contribute to a sustainable future.

7.4.5 Power Conditioning and Controls

Given that FCPS inherently produce low-voltage, high-current outputs, modification through power conditioning is essential for broader application, from mobile transport to residential power supply. Targeting conversions to 400 V DC for automotive applications and 110 V AC for residential energy needs, the system also incorporates mechanisms for excess heat regulation. This comprehensive power conditioning and control framework ensures system adaptability, facilitating real-time adjustments based on diverse parameters including current, voltage, water conductivity, and other critical material properties like pressure, temperature, gas consumption, and humidity.

7.4.6 Balance of Plant

As discussed in 4.3.2, the BoP represents the most sophisticated and resource-intensive segment of the FCS. This microprocessor-managed consortium of subsystems, founded on thermodynamic principles, guarantees consistent FC stack performance amidst varying operational and environmental stresses. Incorporating essential components such as reformers, compressors, pumps, and heat exchangers, the BoP meticulously manages both material and informational flows. Material flows, encompassing gases, coolant, and water, are optimized for thermal efficiency, while informational flows, bridging the "operator–machine" interface, are streamlined through advanced sensor technology.

7.4.7 FCPS Subsystems

Originally discussed in more detail in section 4.3.1 in Chapter 4, a PEM FCPS comprising several essential subsystems is designed to ensure its efficient and effective operation:

- *Fuel Cell Stack Subsystem (FCSS)* encapsulates the FC stack along with all necessary controls to maintain optimal operating conditions. This subsystem interfaces with other components, ensuring a cohesive operation through effective controls and structural hardware.

- *Hydrogen Supply and Conditioning Subsystem (HS&CS)* is tasked with converting and purifying hydrogen-based fuel into a reformate, if necessary. It manages flow metering and the control of pressure, temperature, and mass flow to ensure a steady supply of clean hydrogen.
- *Thermal & Water Management Subsystem (T&WMS)* plays a critical role in sustaining the required operating temperature across the FCPS. It removes excess water from the gas stream and maintains appropriate gas humidification levels for optimal cell performance.
- *Oxidant Supply Subsystem (OSS)* typically includes an air compressor and filters, along with devices to maintain the required pressure, temperature, relative humidity, and mass flow control, ensuring a consistent supply of oxidants to the FCs.
- *Controls & Communications Subsystem (C&CS)* combines hardware and software to oversee all FCPS operations. Using specific algorithms, this subsystem adjusts to various operating conditions, aiming for enhanced performance, durability, and efficiency.
- *Power Conditioner Subsystem (DC/DC or DC/AC Converter)* provides a critical interface between the FC stack output and the electrical grid or application. It ensures that the generated electric power is effectively utilized, matching the required electrical parameters.

An excellent example of an FCPS is the recently developed General Motors HYDROTEC Power Cube (Shown in Fig. 7.16). This fuel cell is capable of efficiently generating up to 77 kilowatts of DC electrical power from energy stored in hydrogen. When combined with other Power Cubes, this technology offers promising and flexible solutions for logistics, transportation, power generation, and other energy-intensive applications.

Fig. 7.16 GM Hydrotec Fuel Cell Power Cube Pre-production. *Source:* GM press materials.

7.4.8 FCPS Functions and Features

The FCPS embodies sophisticated solutions for energy conversion, alongside the management of excess heat and water. Its design integrates an adequate combination of hardware and software across various subsystems, tailoring the FC's functionality to meet specific project objectives. These objectives might include selecting the appropriate fuel, managing emission levels, harnessing rejected heat for additional purposes (such as generating electricity, steam, or heat), and achieving targeted output levels with specific volume or mass efficiency criteria.

The intricate design and operational features of the FCPS are critical for meeting the diverse energy demands of today's applications, ensuring that each system function is supported by a meticulously optimized configuration. This holistic approach to FCS architecture underscores the importance of integrating advanced technological solutions to maximize internal power, efficiency, and economic viability within the evolving landscape of sustainable energy technologies.

7.4.9 FCPS Performance Characteristics

The performance of a FCPS is primarily gauged through its gross and net power outputs. The **gross power** represents the maximal wattage the FC stack can produce under specific operational conditions and load demands. In contrast, **net power** is calculated by deducting the energy consumed by any parasitic loads within the system from the gross power, indicating the actual power available for use.

FCPS efficiency is described as

$$\eta = \frac{P_{\text{electrical out}}}{P_{\text{LHV Hydrogen in}}} \times 100\%$$

where $P_{\text{electrical out}}$ is the electrical power generated by FCPS as a result of electrochemical reaction (41) and is equal to the "net" electrical power at the FCPS load and equal to

$$P_{\text{electrical out}} = V_{\text{Fuel Cell Stack}} \times I_{\text{Load}}$$

$P_{\text{LHV Hydrogen in}}$ is the lower heating value of the hydrogen gas (fuel) prior to reaction and it is equal to

$$P_{\text{LHV } H_2} = 120 \times m_{\text{Avg Hydrogen Flow}}$$

where $m_{\text{Avg Hydrogen Flow}}$ is an average hydrogen mass flow at the FC stack inlet and hydrogen lower heating value is 120 J/g.

A crucial aspect of FCPS functionality is the specification and management of the reactants involved in the electrochemical process. Fuel, typically hydrogen gas, is characterized by precise parameters such as pressure, temperature, mass flow rate, relative humidity, and purity as it enters or exits the fuel stack. These conditions are meticulously controlled to optimize the reaction efficiency and, by extension, the system's overall performance.

Similarly, the oxidizer, generally oxygen from the air, is regulated in terms of pressure, temperature, mass flow rate, and relative humidity to ensure it interacts effectively with the fuel. The electrical power output of an FCPS, a direct result of voltage and current dynamics, varies significantly across different FC types, each with unique requirements for anode gas and oxidant.

FCs are sensitive to impurities such as carbon monoxide (CO), which can poison the cell and degrade its performance. Consequently, selecting suitable catalysts that can withstand the operational temperature ranges and manage oxidants at varying pressures is critical. This nuanced approach to managing reactants and operational conditions underscores the complex interplay of factors that influence the performance and efficiency of FCPS, reflecting the system's adaptability to diverse applications.

7.4.10 Fuel Choice

The architecture and efficiency of an FCPS are significantly influenced by the type of fuel utilized. Historically, the development of fuel systems for automotive applications in the early 1990s prioritized onboard methanol fuel processing, either as a primary fuel choice or as an intermediary step toward the adoption of gasoline. Methanol's preference stemmed from its relative ease of conversion to hydrogen, essential for vehicle FCs.

Methanol undergoes oxidation to carbon monoxide and hydrogen at temperatures below 400 °C and can be steam reformed catalytically at temperatures as low as 250 °C, offering a distinct start-up advantage. It boasts conversion efficiencies exceeding 90%. However, methanol production, primarily from natural gas, demands significant energy, diminishing its well-to-wheel efficiency compared to gasoline. Additionally, methanol's corrosive nature and water affinity complicate fuel distribution, and its acute toxicity surpasses that of gasoline, potentially necessitating additives for safety.

Conversely, gasoline benefits from an established production and distribution network, higher energy density, and lesser toxicity, factors that contribute to enhanced vehicle range. Yet, the conversion of gasoline to hydrogen poses challenges, including high-temperature requirements (around 650 °C), leading to substantial CO, methane, and coke production, alongside increased system complexity.

Other concerns include sulfur and trace metals potentially poisoning the FC stack and complexities in onboard system packaging.

These considerations have gradually shifted the focus toward hydrogen as a preferred fuel source, stored onboard in three primary forms: liquid hydrogen (H_2), compressed H_2, and as metal hydride, each with distinct storage and handling characteristics.

The architecture of gasoline FCSs encapsulates a multifaceted design process tailored to meet a broad spectrum of operational requirements. These include

1) Necessary power output
2) Fuel availability
3) System durability and reliability
4) Cost-effectiveness
5) Power density (both volumetric (kW/L) and gravimetric (kW/kg))
6) System efficiency
7) Start-up time
8) Overall capacity
9) Heat utilization capabilities
10) Number of on/off cycles

Such criteria are pivotal in selecting the appropriate FC technology to ensure optimal functionality and performance.

Current technological developments in FCSs are concentrated across several key areas:

1) DMFC development is prioritized for portable energy solutions, offering convenience for consumer electronics and other small-scale applications.
2) PEMFCs emerge as the primary choice for automotive propulsion and residential power generation, favored for their low operating temperatures and rapid responsiveness to load changes. However, PEMFCs necessitate sophisticated water management systems for humidity control and effective heat rejection strategies. Advances are underway to develop higher temperature membranes to alleviate these issues and enhance carbon monoxide tolerance.
3) SOFCs are explored for distributed power and onsite generation, distinguished by their operation at high temperatures, which eliminates the need for costly precious metals and reduces sensitivity to carbon monoxide.
4) MCFCs and PAFCs are explored for stationary power generation, with PEMFC and SOFC technologies currently leading due to their versatility and efficiency.

PEMFCs, while advantageous for quick load adjustments, face challenges in heat management and humidity control. In contrast, SOFC systems, free from the constraints of precious metal catalysts and less affected by carbon monoxide, present manufacturing complexities and heat loss issues. Notably, integrating SOFCs with gas turbines – capitalizing on the high-temperature exhaust gases – can propel system efficiency to approximately 70% or higher.

Designing FCPS architecture involves navigating trade-offs influenced by cost, efficiency, component power consumption, system complexity, and durability. Project-specific objectives further complicate design processes, encompassing fuel choice, emission standards, heat recovery potential, and desired output specifications.

7.5 The Human Mobility in the 21st Century

As the 21st century progresses, human mobility faces a paradigm shift. The ICE, while a well-established technology, is increasingly recognized for its limitations, particularly concerning natural resource depletion and climate change. This recognition has spurred a search for "green" transportation alternatives, integrating sustainable propulsion methods and advanced technologies such as electric transportation, FCs, and autonomous transportation systems (Fig. 7.17).

7.5.1 Electric Transportation

The exploration of propulsion options reveals a diverse landscape. Electric transportation, leveraging electricity for vehicles ranging from off-road equipment like forklifts to personal mobility devices such as scooters and electric bicycles, represents a significant step toward replacing petroleum-based energy sources. This shift not only reduces greenhouse gas emissions but also diversifies the mobility options available to individuals (Fig. 7.18).

The spectrum of mobility options continues to expand, offering individuals the flexibility to choose between traditional driving, car-sharing services, or embracing autonomous vehicles (AVs). The current reality includes gasoline hybrids and PHEVs, signaling a gradual transition away from ICE. The collective outlook suggests that ICE may remain a preferred option for at least one more

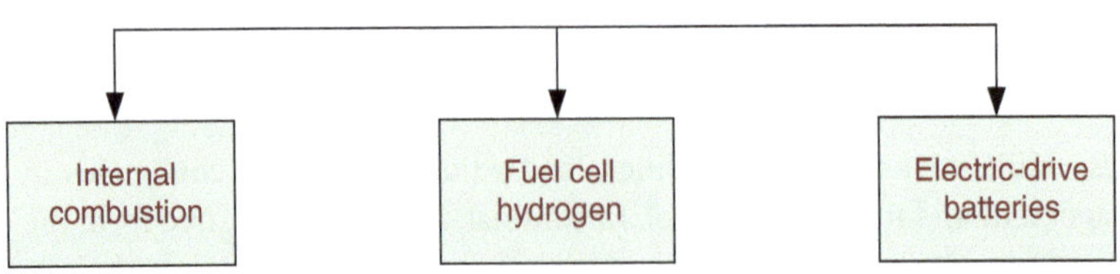

Fig. 7.17 Propulsion Options

Fig. 7.18 The First Electric Vehicle Built in 1897. *Source:* Flickr/CC BY 2.0

human generation, albeit with a steady shift toward hydrogen and electric vehicles as viable and environmentally friendly alternatives (Fig. 7.19).

EVs employ electric motors for propulsion, utilizing either AC or DC motors, each paired with a suitable motor controller to manage power delivery. Initially, EV development favored DC motors, a choice influenced by their origins in the electric forklift industry. DC motors were preferred for their simpler and more cost-effective setup, along with their capability to deliver quick energy bursts essential for acceleration, enhancing vehicle performance.

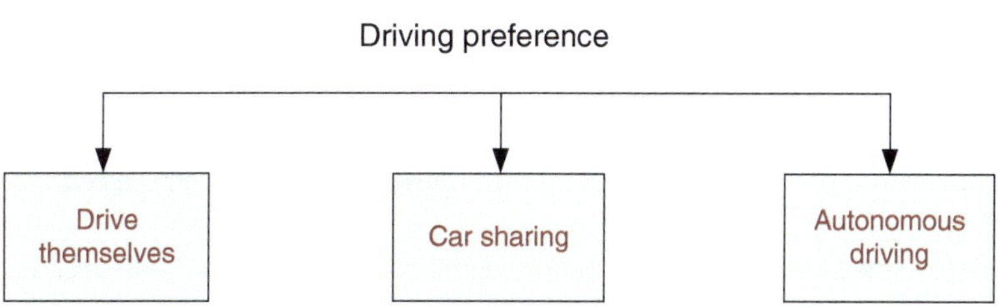

Fig. 7.19 Driving Preference

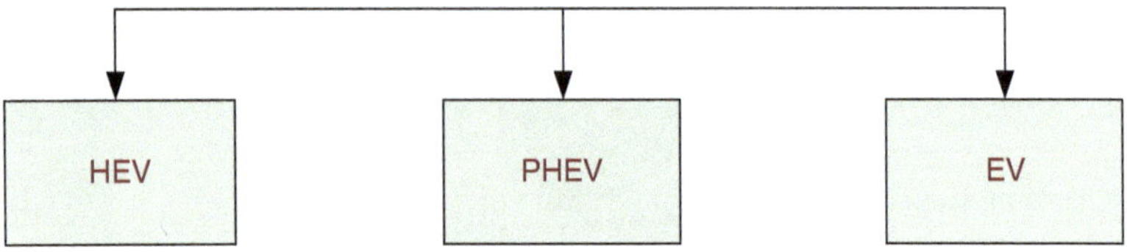

Fig. 7.20 Electric-Drive Options

However, the versatility of AC motors allows the use of nearly any industrial three-phase AC motor, though the industry is now advancing toward motors specifically designed to meet EV requirements. These requirements include considerations for weight, configuration, packaging, and electrical parameters. A significant advantage of AC motors and their controllers is the ability to regenerate energy. Known as "regen," this feature enables the capture of energy during an EV's deceleration, either by using the drive motor as a generator or by operating a dedicated generator. This regenerative braking process returns energy to the battery pack, marginally increasing the driving range.

Figure 7.20 illustrates that all-electric drive vehicles fall into three primary categories: HEVs, PHEVs, and EVs. These classifications highlight the use of electricity either as the sole energy source or to augment the efficiency of conventional vehicle designs, marking a significant step toward sustainable transportation solutions.

7.5.2 Hybrid Electric Vehicles

HEVs stand at the forefront of automotive innovation, integrating two distinct propulsion sources: an ICE that operates on petroleum-based fuels or alternative fuels, and an electric motor powered by energy stored in a battery. This dual-source system allows HEVs to optimize fuel efficiency and reduce emissions compared to their conventional counterparts.

One of the key features of HEVs is the ability to charge the battery through regenerative braking and the ICE itself. This innovative charging mechanism captures energy typically lost during braking by utilizing the electric motor as a generator to store energy back into the battery. Additionally, the electric motor provides supplemental power, supporting auxiliary vehicle loads such as audio systems and headlights, and even enabling the engine to idle less when the vehicle is stationary.

HEVs are designed to enhance fuel economy significantly. Some models can operate on electric power alone for short distances (40–60 miles) at low speeds, further contributing to their efficiency and reduced environmental impact. Unlike PHEVs,

HEVs do not require external charging sources since they cannot be plugged in; instead, they rely solely on regenerative braking and the ICE for battery charging.

7.5.3 Plug-in Hybrid Electric Vehicles

PHEVs represent a sophisticated bridge between HEVs and fully EVs, incorporating both an ICE and an electric motor. The defining feature of PHEVs is their ability to be recharged from external electrical sources, unlike HEVs that rely solely on regenerative braking and the ICE for battery charging. This capability allows some PHEVs to travel up to 70 miles purely on electric power, offering a versatile driving experience that can switch to gasoline as needed, akin to conventional hybrids.

PHEVs are equipped with larger battery packs than their HEV counterparts, enabling an all-electric driving range that exceeds 40 miles for light-duty models. This extended range is particularly beneficial for urban driving, where vehicles can operate mostly on electricity, significantly reducing tailpipe emissions. Once the battery's charge is depleted, the vehicle seamlessly transitions to the ICE, ensuring uninterrupted mobility. The dual-mode operation ensures that, even when relying on gasoline, PHEVs consume less fuel and produce fewer emissions compared to traditional vehicles.

The environmental benefits of PHEVs are twofold. Primarily, when operating on battery power, these vehicles emit no tailpipe pollutants, directly contributing to cleaner air quality. Secondly, even when the ICE is in use, PHEVs maintain higher fuel efficiency and lower emissions than conventional vehicles. The extent of these benefits largely depends on the frequency of charging and the distances traveled; regular charging to leverage the electric driving mode maximizes the environmental advantages.

Some PHEVs, known as extended-range electric vehicles (EREVs), further blur the lines between HEVs and fully EVs by offering substantial electric-only ranges. This flexibility ensures that drivers can enjoy the benefits of electric driving for daily commutes while retaining the ICE for longer trips, providing a comprehensive solution to current limitations in EV charging infrastructure.

7.5.4 All-Electric Vehicles

EVsrepresent a significant shift in automotive technology, utilizing a single internal source for propulsion: a battery that stores electrical energy to power an electric motor. EVs are recharged by connecting to an external electrical power source, marking a departure from traditional fuel-dependent vehicles.

EVs are designed to operate solely on electricity, utilizing one or more electric motors powered by rechargeable battery packs. The configuration of motors varies, with some EVs featuring a single motor, while others deploy dual motors for enhanced performance on front and rear wheels, or individual motors for each wheel, optimizing torque distribution and vehicle dynamics.

7.5.5 Comparison with ICEs

EVs offer several advantages over their ICE counterparts:

- *Energy Efficiency*: EVs boast an efficiency rate of about 60% in converting electrical energy from the grid to power at the wheels, significantly outperforming gasoline vehicles, which convert only about 21% of the energy stored in gasoline.
- *Environmental Impact*: With zero tailpipe emissions, EVs offer a cleaner alternative to ICE vehicles. While electricity generation from fossil fuels still poses environmental challenges, renewable sources like hydro, solar, and wind contribute to a significantly reduced carbon footprint.
- *Performance and Maintenance*: Electric motors ensure quiet, smooth operation, and quicker acceleration, coupled with lower maintenance costs due to fewer moving parts and reduced wear.
- *Durability*: Although direct comparison data is limited, the minimal friction in EV components suggests a potential for a longer lifespan compared to ICE vehicles.
- *Infrastructure Compatibility*: EV charging infrastructure leverages existing electrical networks, presenting a more straightforward implementation than the development required for hydrogen fuel distribution.

7.5.6 Fuel Cells vs. Electric Batteries

The future of electric mobility may hinge on the competition between electrochemical propulsion from onboard hydrogen FCs and energy storage using batteries. FCs offer a technically feasible and economically viable option, yet the current lack of hydrogen production, efficient storage solutions, and distribution infrastructure remains a barrier. Conversely, battery technology, while logistically simpler and less infrastructure-intensive, faces challenges in energy density, driving range, recharge times, weight, and cost.

The advancements in addressing these challenges will shape the future of EVs and the evolving landscape of autonomous transportation, balancing the trade-offs between FC and battery technologies to determine the most sustainable and efficient path forward for electric mobility (Fig. 7.21).

Over the past two centuries, the automotive industry has embarked on a relentless pursuit of innovation to deliver environmentally friendly and affordable transportation solutions. This journey has seen the introduction of groundbreaking technologies such as direct injection, catalytic converters, electric propulsion through batteries, and electrochemical engines powered by hydrogen. These

Fig. 7.21 Self-Driving Car with HUD (Heads Up Display) Mockup. *Source:* scharfsinn86/ Adobe Stock Photos

advancements have significantly reduced the environmental impact of vehicles, marking a pivotal shift toward sustainability in transportation.

In recent years, we are witnessing an unprecedented revolutionary pace in the development of autonomous technology. The iconic American thruways and the automobile, once symbols of America's identity, showcasing individual status and interests through choices between SUVs, convertibles, minivans, or sports cars, are poised for a transformative shift. This traditional image of personal mobility is on the cusp of becoming a mere memory.

Having spent a substantial portion of my career in the automotive industry, I once viewed AVs and driverless cars as mere science fiction, a distant future for generations to come. However, this perspective has dramatically changed. The rapid advances in artificial intelligence, coupled with breakthroughs in Big Data analysis, sensors, and actuators, have made autonomous driving not just a possibility but a reality currently in testing by numerous manufacturers. This leap forward is reshaping our understanding of mobility, promising a future where vehicles navigate the world with unprecedented efficiency and safety, heralding a new era in the automotive industry.

7.5.7 Social Influence of Autonomous Cars

The advent of autonomous cars promises to revolutionize automobile transportation, potentially shifting the concept of car ownership from personal pride and pleasure to a commoditized service, akin to phones or other electronic gadgets, utilized purely on demand. This transition challenges two centuries of automotive history and the mass production business model pioneered by Ford, potentially marking the end of traditional vehicle ownership and dealer–customer relationships. In this new paradigm, ownership may consolidate under large corporations dominating the driverless landscape, transforming the automotive industry's structure.

AVs herald unprecedented opportunities for social welfare enhancement. Mobility for the elderly and disabled populations is set to improve significantly, making transportation more accessible and affordable for those living on a budget. The potential benefits extend to saving lives by reducing accidents, alleviating traffic congestion, conserving natural resources, and minimizing environmental pollution. Additionally, the need for fewer vehicles could lead to more efficient land use and an increase in mobility for disabled individuals. Importantly, the time saved from driving could be redirected toward human creativity and enjoyment.

The shift toward autonomous driving necessitates increased electrical energy production and may require modifications to transmission lines or the development of wireless electricity distribution systems. The establishment of electric recharge stations and battery pack exchange facilities will become essential, alongside a probable decrease in petroleum production.

This technical revolution calls for new government legislation regarding cyber security, novel financial models for vehicle acquisition, comprehensive changes in car insurance, and the creation of new employment opportunities. The insurance industry, in particular, may undergo significant changes, with policies transitioning from individual owners to encompass users and service providers of AVs.

The rise of autonomous cars is set to transform not only how we view and use vehicles but also impact social welfare, environmental conservation, and the economic and legislative frameworks governing transportation. As we stand on the brink of this transformation, the collective challenge will be to navigate these changes responsibly, ensuring benefits are maximized while mitigating potential downsides.

7.5.8 Employment and AVs

The advent of AV technology brings to the forefront significant employment challenges and the potential for societal disruption. A key concern is the future of the workforce previously engaged in sectors directly impacted by automation, such as transportation repair, truck driving, infrastructure development, vehicle production, and related financial services.

The transition to AVs will likely result in

- Shifts in small business operations, including convenience stores, taxi services, automotive repair, and body shops, potentially leading to the closure of businesses or the need to adapt to new market realities.

- Creation of lower-paid positions for dispatchers tasked with managing autonomous fleets.
- Reduced labor demand in agricultural production, automotive manufacturing, and the construction industry due to increased efficiency and automation.
- Evolving legal frameworks and legislation to accommodate autonomous technology, impacting legal professions and law enforcement.
- A decline in petroleum production as electric AVs gain prominence.

Despite these challenges, history demonstrates society's capacity to adapt to technological advancements, transitioning from horse-drawn carriages to automobiles and from manual labor to machine-assisted tasks. The shift toward AVs is expected to generate new employment opportunities in the automotive IT sector, cybersecurity, Big Data analysis, and the production of sensors and actuators. This transition necessitates a strategic approach to workforce development and retraining programs to equip displaced workers with the skills needed for emerging roles in a technologically driven economy.

7.5.9 Urban Future in the Era of AVs

The emergence of AVs is set to radically transform urban landscapes, dictating new requirements for city planning and infrastructure. This shift will necessitate comprehensive changes, including the total redesign of parking lots, street layouts for enhanced pedestrian safety, and the integration of new vehicle technologies.

7.5.9.1 Key Transformations Induced by AVs

1) *Redesign of Urban Spaces*: The conventional design of parking lots and street planning will undergo significant changes to accommodate AVs, ensuring safe and efficient interaction with pedestrians and new generations of vehicles.
2) *Integrated Traffic Management*: The adoption of peer-to-peer networks between AVs and smart intersections will become crucial for regulating traffic flow, minimizing congestion, and improving safety.
3) *Impact on Public Transportation*: Autonomous cars and buses are poised to redefine commuter systems, potentially replacing current bus and rail networks. This evolution could alter the economics of city planning, affecting traditional sources of funding like speeding tickets and parking meters.
4) *Shift in Vehicle Ownership*: With the decline in individual ownership, fleets of unmanned vehicles will become prevalent, raising questions about fleet management and operation. Parking needs will transition to communal spaces for AVs, and traditional gasoline stations will be replaced by centralized electric and hydrogen refueling stations.

7.5.9.2 Social and Economic Implications

1) *Limited Personal Transportation*: Inside cities, personal transportation may become unnecessary or very limited, especially for long-distance travel.
2) *Architectural Changes*: The disappearance of home garages may influence residential architecture and reduce dwelling costs.
3) *Reduction in Repair Centers*: A decrease in the number of repair centers could lower the expenses associated with car maintenance.
4) *Enhanced Public Transportation Utilization*: The effective use of public transportation by AVs could lead to redesigned streets and improved safety measures.
5) *Infrastructure Modifications*: Changes will be necessary for energy delivery systems and urban land usage to support AVs.
6) *Creation of Charging Stations*: The development of charging stations will become essential for supporting electric AVs.
7) *Traffic Congestion Reduction*: Big Data utilization by AVs promises to minimize traffic congestion, improving overall transportation efficiency.
8) *Lowered Delivery Costs*: The automation of delivery services may reduce costs and labor requirements in various sectors, including agriculture.

7.5.9.3 Transportation Efficiency

1) *Increased Vehicle Usage Efficiency*: AVs will likely spend less time parked, optimizing the use of vehicles.
2) *Maintenance Cost Reduction*: The need for personal and commercial vehicle maintenance could decrease, leading to new business models for original equipment manufacturer (OEM) production, service, and distribution.
3) *Streamlined Dealerships*: The number of dealerships might be limited, impacting automotive manufacturers and suppliers.
4) *Gasoline Station Decline*: A substantial decrease in the number of gasoline stations is anticipated, alongside more efficient truck delivery usage.
5) *Ecological Improvements*: The shift toward AVs is expected to bring significant environmental benefits.

7.5.10 Financial Implications of AVs

The advent of AVs will radically alter the financial landscape of the automotive industry:

1) *Automotive Loans*: The shift toward AVs and service-based transportation could diminish the need for individual automotive loans, as personal vehicle ownership declines.
2) *Automotive Insurance*: New strategies will be required to address the unique risks and liabilities associated with AVs, potentially leading to the development of innovative insurance products.

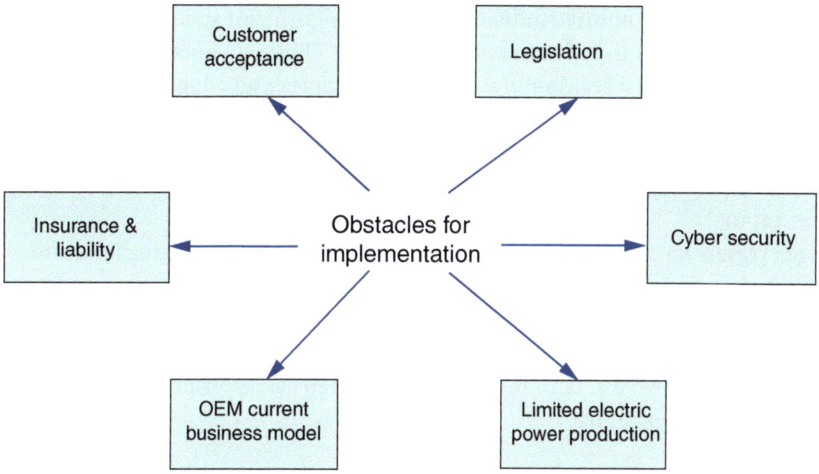

Fig. 7.22 Obstacles for Implementation

3) *Dealership and OEM Policies*: Traditional dealership models and OEM strategies may need to evolve to accommodate the changing dynamics of vehicle sales and service in an AV-dominated market.
4) *Medical Cost Compensation*: The anticipated reduction in traffic accidents due to AVs could significantly impact medical cost compensation, necessitating a redistribution of medical resources (Fig. 7.22).

The path to widespread AV adoption is not solely defined by technological advancements. Regulatory, liability, and legislative challenges represent substantial hurdles. While technological robustness continues to improve through extensive testing, achieving a regulatory framework that supports the safe integration of AVs into public roads is paramount. According to Navigant Research, the global market for vehicles with Level 2–4 autonomy is expected to reach 85 million annually by 2035, underscoring the pressing need to address these nontechnological barriers.

7.5.11 Self-Driving Functionality and Levels of Autonomy

The evolution of AV technology is systematically categorized by the Society of Automotive Engineers (SAE) into 6 levels of complexity, ranging from Level 0, where the human driver is in complete control of the vehicle, to Level 5, which represents full automation. At Level 5, all driving tasks are managed by an AI computerized system, capable of making decisions and reacting to traffic conditions more rapidly and accurately than a human driver. This level of autonomy ensures seamless vehicle operation in diverse scenarios, from routine commutes to complex, life-threatening traffic situations.

Currently, the automotive industry has made significant strides toward integrating semiautonomous features into vehicles. These advancements include automatic emergency braking in the face of obstacles and adaptive cruise control for highway driving. Such features represent early iterations of autonomy, laying the groundwork for more sophisticated systems that promise to navigate traffic jams and other challenging driving conditions with minimal or no driver input.

Major players across the transportation and service sectors, including manufacturers like Tesla, BMW, Volvo, Audi, and Ford, alongside tech companies such as Uber and Waymo, are at the forefront of developing and testing vehicles with Level 4 and 5 autonomy. These efforts indicate a strong industry commitment to transitioning toward fully AVs, with the potential for their widespread deployment on public roads in near future.

7.5.12 Safety Considerations in AV Technology

The advent of AVs brings to the forefront critical safety considerations that must be addressed to ensure their successful integration into daily transportation:

- *Accidents and Legal Responsibility*: As AVs navigate driverless transportation, determining legal responsibility in the event of accidents becomes complex. This necessitates clear legislation and regulatory frameworks to address liability issues in incidents involving AVs.
- *Cybersecurity*: Protecting AVs from cyber threats is paramount to prevent potential criminal or terrorist activities. Establishing robust cyber safety security measures is essential to safeguard the integrity and functionality of autonomous transportation systems.

7.5.13 Cloud Network Database Establishment for AVs

The transition from traditional automotive design metrics, such as valves and piston configurations, to electric and AVs underscores the importance of developing a cloud-based network. This network is crucial for aggregating and analyzing vast amounts of data from multiple sources, including vehicle location, car-to-car communications, road conditions, weather, traffic flow, vehicle internals, and street sensors. Overcoming the challenge of processing this data swiftly for moving vehicles is pivotal for real-time decision-making.

This cloud network transforms the vehicle grid into a dynamic information ecosystem, facilitating data-driven decisions on traffic management, safety protocols, and individual vehicle maneuvers. While automotive manufacturers focus on vehicle production, service companies like Google, Uber, and Lyft are at the forefront of managing Big Data, essential for enhancing autonomous driving capabilities.

7.5.14 The Rise of AVs

The once science-fictional dream of robotic cars cruising our streets is now an undeniable reality. Today's technological advancements have paved the way for vehicles to navigate reliably through ever-changing environments, connect seamlessly, and ensure security through sophisticated cloud networks. This new era of technology demands the development of advanced software and computerized systems dedicated to autonomous driving decision-making. These systems leverage cloud architecture and state-of-the-art hardware devices, including LIDAR (light detection and ranging) sensors, which utilize lasers to create detailed 3D representations of the surroundings, among other navigation and communication technologies.

As we witness an expansion in autonomous driving capabilities within current car models, there is a growing expectation that a diverse range of robotic vehicles will enter production within the next decade. This shift is poised to significantly transform human mobility, leading to a decrease in personal car ownership, an increase in short-term vehicle rentals for individual needs, and a reduction in workforce requirements within the transportation sector. The evolution toward AVs heralds a new chapter in how we perceive and engage with personal and public transportation, marking a fundamental shift in societal norms and economic structures.

7.6 Conclusion

As we stand on the precipice of global change, the future well-being of our planet and its burgeoning population of over 8 billion souls hinges not merely on food security but critically on our ability to curtail the consumption of natural energy resources. This book has ventured to outline the contours of a prospective hydrogen-based society, a vision teeming with promise for addressing some of the most pressing challenges of our time. Such a society heralds numerous advantages:

- *Limited greenhouse gas emissions*, significantly mitigating climate change.
- *Conservation of fossil fuel resources*, safeguarding them for future generations.
- *Transformation of intermittent renewable sources into hydrogen*, providing a stable energy supply.
- *Advanced hydrogen storage solutions*, offering alternatives to conventional battery storage.

- *A driverless future in FC transportation*, precipitating shifts in employment landscapes and urban planning.
- *Enhanced energy conversion efficiency*, optimizing the use of available resources.
- *Localized power generation*, reducing dependency on centralized energy networks.
- *Introduction of wireless energy distribution*, revolutionizing how we access and share power.
- *Economic stability and prosperity*, underpinning global development and cooperation.

Comparing hydrogen technology to our existing, predominantly wired energy distribution systems reveals its potential to revolutionize power generation and consumption. The current, centralized model, reliant on extensive transmission networks, faces challenges in cost and efficiency. However, the societal shift toward distributed power generation, paralleled by the transition from landlines to mobile communications, underscores a growing preference for decentralization.

Envisioning a future where FCVs not only transport us but also contribute electricity back to the grid while parked, we glimpse a paradigm of decentralized power production. This model envisages every household as both a consumer and producer of energy, leveraging renewable sources to feed into a Hydrogen Energy Web, akin to the internet's transformative impact on information exchange.

For this hydrogen society to come to fruition, a concerted effort from governmental bodies, alongside the ingenuity of scientists and engineers, is imperative. The transition calls for a collaborative approach to energy production, mirroring the participatory ethos of the digital age.

This book aspires not merely to inform but to inspire. It invites you to participate in the journey toward a hydrogen society, a venture that promises not just environmental sustainability but a reimagined economic and social landscape for future generations.

Used Literature

Chapter 01

1 https://www.iea.org/reports/global-hydrogen-review-2022/executive-summary.
2 EUR-Lex – 52020DC0301 – EN.
3 https://www.worldenergy.org/assets/downloads/PUB_Survey-of-Energy-Resources_2004_WEC.pdf.
4 https://www.eia.gov/outlooks/steo/tables/pdf/3btab.pdf.
5 https://www.bp.com/content/dam/bp/business-sites/en/global/corporate/pdfs/energy-economics/energy-outlook/bp-energy-outlook-2024.pdf.
6 worldenergy.org/assets/images/imported/2013/09/Complete_WER_2013_Survey.pdf.
7 Prinzhofer, A., Cissé, T., Sidy, C., and Diallo, A.B. (2018). Discovery of a large accumulation of natural hydrogen in Bourakebougou (Mali). *International Journal of Hydrogen Energy* 43 (42): 19315–19326. https://doi.org/10.1016/j.ijhydene.2018.08.193. S2CID 105839304.
8 https://www.renewableenergyworld.com/storage/.
9 https://www.youtube.com/watch?v=3HxdJzJLqj8.
10 https://earth.org/data_visualization/a-brief-history-of-co2/.
11 https://www.reuters.com/article/factcheck-volcanoes-co2/fact-check-volcanoes-do-not-produce-more-co2-emissions-than-human-activity-idUSL1N2XV1HA.
12 https://www.theguardian.com/environment/2019/nov/05/climate-crisis-11000-scientists-warn-of-untold-suffering.
13 https://www.britannica.co/precision.
14 climate.nasa.gov.
15 https://earth.org/data_visualization/a-brief-history-of-co2/.
16 https://tuchiki.livejournal.com/106638.html.

Hydrogen Energetics: A Future Clean Energy Carrier, First Edition. Roman J. Press.

17 https://energy.ec.europa.eu/topics/energy-systems-integration/hydrogen_en.
18 https://www.rechargenews.com/energy-transition/europe-is-never-going-to-be-capable-of-producing-its-own-hydrogen-in-sufficient-quantities-eu-climate-chief/2-1-1212963.
19 https://www.iea.org/reports/global-hydrogen-review-2022/executive-summary.
20 http://www.cnn.com/2003/ALLPOLITICS/02/06/bush-energy/index.mhtl (accessed 25 May 2017).
21 US Government Energy Information Administration. http://www.eia.doe.gov, http://wilcoxen.maxwell.insightworks.com/pages/804.html (accessed 25 May 2017).
22 What is tidal energy? http://www.oceanenergycouncil.com/ocean-energy/tiday-energy (accessed September 2017).

Chapter 02

1 https://www.iea.org/reports/wind-electricity.
2 https://www.youtube.com/watch?v=_nMZK38i4q0.
3 Aeromine. http://www.aerominetechnologies.com/news/
4 https://investingnews.com/top-natural-gas-producers.
5 https://www.ocean-energy-systems.org/ocean-energy/what-is-ocean-energy.
6 https://www.eia.gov/energyexplained/hydropower/tidal-power.php.
7 https://education.nationalgeographic.org/resource/tidal-energy.
8 https://www.eia.gov/tools/faqs/faq.php?id=427&t=3.
9 www.teslasociety.com.
10 https://www.iea.org/energy-system/renewables/wind.
11 https://www.bbc.com/news/science-environment-63950962.
12 OPEC. (2022). Annual Statistical Bulletin.
13 US Energy Information Administration. http://www.eia.doe.gov.
14 https://investingnews.com/top-natural-gas-producers.
15 Douglas, R.B., Toman, M.A. (2015). Analyzing Nonrenewable Resource Supply.
16 https://en.wikipedia.org/wiki/Coal.
17 https://en.wikipedia.org/wiki/Hubbert_curve.
18 https://www.eia.gov/energyexplained/nuclear.
19 https://www.eia.gov/tools/faqs/faq.php?id=427&t=3.
20 https://en.wikipedia.org/wiki/Sedimentary_rock.
21 https://www.eia.gov/international/data/world.

Chapter 03

1 https://en.wikipedia.org/wiki/Hydrogen_bond.
2 Wigner, E. and Huntington, H.B. (1935). On the possibility of a metallic modification of hydrogen. *Journal of Chemical Physics* 3 (12): 764. https://doi.org/10.1063/1.1749590.
3 Guillot, T., Stevenson, D.J., Hubbard, W.B., and Saumon, D. (2004). Chapter 3: The interior of jupiter. In: *Jupiter: The Planet, Satellites and Magnetosphere* (ed. F. Bagenal, T.E. Dowling, and W.B. McKinnon). Cambridge University Press.
4 https://www2.lbl.gov/abc/wallchart/chapters/02/3.html.
5 https://www.britannica.com/science/hydrogen.
6 Takacs, G.A., Bailey, A.V., Miri, M.J., and Santhanam, K.S.V. *Introduction to Hydrogen Technology*, 2. Roman J. Press.
7 Cammack, R., Frey, M., and Robson, R. (2001). *Hydrogen as Fuel: Learning from Nature*. London: Taylor and Francis.
8 Rigden, J.S. (2002). *Hydrogen: The Essential Element*. Cambridge: MA Harward University Press.
9 Dicks, A.L. and Rand, D.A.J. *Fuel Cell Systems Explained*, 3e.
10 Barron, A. (2020). *Chemistry of the Main Group Elements*. MiDAS Green Innovations 9781838008512.
11 Ciniviz, M. and Köse, H. (2012). Hydrogen use in internal combustion engine: a review. *International Journal of Automotive Engineering and Technologies* 1 (1): 1–15.
12 John Wilson and Griffin Burgh (2003). *The Hydrogen Report*. TMG/The Management.
13 Zuttel, A. and Orimo, C. (2002). Hydrogen in nanostructured, carbon-related, and metallic mate-rials. *MRS Bulletin* 27 (9): 705–710.
14 Bowman, R.C. and Fultz, B. Jr. (2002). Metallic hydrides I: hydrogen storage and other gas-phase applications. *MRS Bulletin* 27 (9): 688–691.
15 Dillon, A.C., Gilbert, K.E.H., Parilla, P.A., et al. (2002). Hydrogen Storage in Carbon Single-Wall Nanotubes, National Renewable Energy Laboratory Golden, CO 80401-3393 Proceedings of the 2002 U.S. DOE Hydrogen Program Review NREL/CP-61032405.
16 Adsorption of Hydrogen Molecules on Carbon Nanotubes http://pubs.acs.org/doi/abs/10.1021/acs.jpca.5b12574 (accessed 28 July 2016)
17 http://en.wikipedia.org/wiki/Hydride (accessed 10 June 2017).
18 http://www.nrel.gov/hydrogen/proj_storage.html (accessed 10 June 2017).

Chapter 04

1 https://www.energy.gov/eere/fuelcells/types-fuel-cells.
2 https://www.energy.gov/eere/fuelcells/articles/comparison-fuel-cell-technologies-fact-sheet.
3 https://www.energy.gov/eere/fuelcells/articles/us-department-energy-hydrogen-and-fuel-cell-technologies-office-and-globalinfocost.
4 www.fuelcellenergy.com.
5 https://www.sciencedirect.com/topics/chemistry/direct-methanol-fuel-cell.
6 https://www.energy.gov/eere/fuelcells/fuel-cell-systemsprocessor.
7 https://www.energy.gov/eere/fuelcells/parts-fuel-cell#bipolar-plates.
8 Nelson, D. (2003). *SAE Seminars "Automotive Fuel Cell Systems"*. Virginia Tech.
9 https://en.wikipedia.org/wiki/Membraneless_Fuel_Cells.
10 https://www.sciencedirect.com/science/article/abs/pii/S0360319911001066e.
11 Dicks, A.L. and Rand, D.A.J. *Fuel Cell Systems Explained*, 3e. Wiley.
12 Santhanam, K.S.V., Press, R.J., Miri, M.J. et al. (2018). *Introduction to Hydrogen Technology*, Second Edition, Wiley.

Chapter 05

1 https://www.energy.gov/eere/fuelcells/safety-codes-and-standards.
2 https://www.aiche.org/sites/default/files/community/278666/aiche-community-site-newsletter/2984996/3globalh2safetycodesnstandardsdrchrislafleurpe 21jul2021.pdf.
3 https://www.osha.gov/laws-regs/regulations/standardnumber/1910/1910.103.
4 https://www.nfpa.org/codes-and-standards/all-codes-and-standards/list-of-codes-and-standards.
5 https://www.sandia.gov/labnews/2022/08/25/improving-hydrogen-safety-codes-and-standards.
6 https://www.aiche.org/sites/default/files/community/278666/aiche-community-site-newsletter/2984996/3globalh2safetycodesnstandardsdrchrislafleurpe21 jul2021.pdf.
7 https://inis.iaea.org/collection/NCLCollectionStore/_Public/38/117/38117074.pdf.
8 http://www.eere.energy.gov/hydrogenandfuelcells/codes/.
9 http://en.wikipedia.org/wiki/Hydrogen_station.
10 https://www.sciencedirect.com/science/article/abs/pii/S0925400511003674.
11 Eliasson, B. and Bossel, U. (2003). The Future of the Hydrogen Economy: Bright or Bleak?
12 http://en.wikipedia.org/wiki/Hydrogen_economy.

13 Hübert, T., Boon-Brett, L., Black, G., and Banach, U. (2011). Hydrogen sensors – a review. *Sensors and Actuators B: Chemical* 157 (2): 329–352. https://doi.org/10.1016/j.snb.2011.04.070.

14 Kaidarova, A. and Kosel, J. Physical sensors based on laser-induced graphene: a review. *IEEE Sensors Journal* 21 (11): 12426–12443.

15 https://www.e-asct.org/journal/view.html?volume=31&number=4&spage=79.

16 https://www.crowcon.com/blog/how-long-will-my-gas-sensor-last/sensors.

17 https://en.wikipedia.org/wiki/Flow_battery.

18 http://geology.about.com/library/weekly/aa080402a.htm.

19 http://www.worldwatch.org/features/consumption/sow/trendsfacts/2004/07/7.

20 Crutzen, P.J. (2000). *Source: IGBP Newsletter 41, May*. Max-Planck-Institute for Chemistry Division of Atmospheric Chemistry.

21 Stoermer, E.F. (n.d.). *Center for Great Lakes and Aquatic Sciences*. Ann Arbor, Michigan, USA: University of Michigan.

22 https://ieeexplore.ieee.org/abstract/document/4450599.

23 https://fardapaper.ir/mohavaha/uploads/2017/09/212245484512615484121 51.pdf.

24 https://essinc.com/.

25 https://www.sciencedirect.com/topics/engineering/energy-storage-system.

26 https://www.greentechmedia.com/articles/read/.most-promising-long-duration-storage-technologies-left-standing.

27 https://www.carboncollective.co/sustainable-investing/battery-energy-storage-systems-bess.

28 https://en.wikipedia.org/wiki/Energy_storage.

29 https://www.lyellcollection.org/doi/10.1144/SP528-2022-88.

30 https://www.virta.global/vehicle-to-grid-v2g#three.

31 Schlapbach, L. (2011). Hydrogen as a fuel and its storage for mobility and transport. *MRS Bulletin* 27: 675–679.

32 Dagani, R. (2002). Tempest in tiny tube. Chemical Engineering News, January 14.

33 Zuttel, A. and Orimo, C.-i. (2002). Hydrogen in nanostructured, carbon-related, and metallic materials. *MRS Bulletin* 27 (9): 705–710.

34 Robert C. Bowman Jr and Brent Fultz, Metallic hydrides I: hydrogen storage and other gas-phase applications, *MRS Bulletin*, 2002, 27, 9, p.688–691.

35 Anne C. Dillon, M.J. Heben, T. Gennett, and P.A. Parilla (2002). Storage in Carbon Single-Wall Nanotubes.

36 http://www.whitehouse.gov/infocus/energy/.

37 http://en.wikipedia.org/wiki/Hydride.

38 https://en.wikipedia.org/wiki/Powerpaste#cite_note-FraunhoferPowerpaste2019-1.

39 https://www.sciencedirect.com/science/article/pii/S1002007121000265.

40 https://www.sciencedirect.com/book/9780128194768/nanomaterials-for-hydrogen-storage-applications.

41 https://www.nrel.gov/docs/fy02osti/32405b28.pdf.
42 https://www.energy.gov/eere/fuelcells/metal-hydride-storage-materials.
43 Cammack, R., Frey, M., and Robson, R. (2001). *Hydrogen as Fuel: Learning from Nature*. London: Taylor and Francis.
44 Moore, R.B. and Raman, V. (1998). Infrastructure for fuel cell transportation. *International Journal of Hydrogen Energy* 23 (7): 617–620.
45 http://www.methanol.org/pdfFrame.cfm?pdf=HydrogenEconomyReport2003.pdfHydrogen.
46 http://en.wikipedia.org/wiki/Electrolysis.
47 https://www.energy.gov/eere/fuelcells/hydrogen-pipelines.
48 https://www.greentechmedia.com/articles/read/green-hydrogen-in-natural-gas-pipelines-decarbonization-solution-or-pipe-dream piping.
49 Wilson, J. and Burgh, G. (n.d.). Energizing Our Future: Rational Choices for the 21st Century
50 Melaina, M.W., Antonia, O., and Penev, M. ((n.d.).). *Blending Hydrogen into Natural Gas Pipeline Networks: A Review of Key Issues*. NREL.
51 https://en.wikipedia.org/wiki/Nord_Stream_1.
52 https://byjus.com/jee/hydrogen-bonding/.
53 Nejat Veziroglu, T. ((n.d.).). *Hydrogen Energy System: A Permanent Solution to Global Problems*. University of Miami.

Chapter 06

1 https://link.springer.com/article/10.1007/s40789-018-0196-3.
2 https://www.sciencedirect.com/science/article/abs/pii/0360319983901751.
3 Hallenbeck, P.C. and Benemann, J.R. (2002). Biological hydrogen production: fundamentals and limiting processes. *International Journal of Hydrogen Energy* 27 (11–12): 1185–1193.
4 Ohi, J. (2005). Hydrogen energy cycle: an overview. *Journal of Materials Research* 20 (12): 3180–3187.
5 Scott, D.S. (2004). For better or worse. *International Journal of Hydrogen Energy* 29: 449.
6 Steinberg, M. (1999). Fossil fuel decarbonization technology for mitigating global warming. *International Journal of Hydrogen Energy* 24: 771–777.
7 https://pubs.acs.org/doi/10.1021/acs.iecr.1c01679.
8 https://www.researchgate.net/publication/223165924_Hydrogen_Via_Methane_Decomposition_An_Application_for_Decarbonization_of_Fossil_Fuels.
9 http://large.stanford.edu/courses/2011/ph240/nagasawa1/docs/038409.pdf.
10 U.S. Department of Energy. (n.d.) *Energy Efficiency and Renewable Energy, Clean Cities Fact.*

11 Magrini-Bair, K.A., Czernik, S., French, R., and Chornet, E. (2003). *Fluidizable Catalysts for Hydrogen Production from Biomass Pyrolysis/Steam Reforming*. FY 2003 Progress Report,. National Renewable Energy Laboratory.

12 Czernik, S., French, R., Evans, R., and Chornet, E. (2003). Hydrogen from post-consumer residues. *US DOE Hydrogen and Fuel Cells Merit Review Meeting, Berkeley, CA*, pp. 1–17.

13 Evans, R., Boyd, L., Elam, C. et al. (2003). Hydrogen from biomass-catalytic reforming of pyrolysis vapors. *US DOE Hydrogen, Fuel Cells & Infrastructure Technologies Program – 2003 Annual Merit Review Meeting*.

14 U.S. Environmental Protection Agency. (n.d.). *Landfill Methane Outreach Program*.

15 Energy Efficiency and Conservation Authority, New Zealand. *Biogas and Landfill Gas*. http://www.eeca.govt.nz/pdfs/biogas_and_Landfill_gas_fact_sheet.

16 *Landfill Gas Utilization at RIT's Campus Feasibility Study*. PON No. 732-02 RIT March 2005.

17 *QuestAir Technologies PSA Technical Information*. http://questair.gssiwebs.com/applications/biogas.htm.

18 Turner, J. (2003). *Photoelectrochemical Water Splitting*.

19 Varner, K., Warren, S. and Turner, J.A.,(2002). Photoelectrochemical systems for hydrogen production. *Proceedings of the 2002 US DOE Hydrogen Program Review*. NREL/CP-610-32405.

20 Ghirardi, M. and Seibert, M. (2003). *Algal Hydrogen Photoproduction*.

21 Dahl, J. et al. (2003). *High Temperature Solar Splitting of Methane to Hydrogen and Carbon*.

22 https://www.researchgate.net/publication/221950172_Solar_hydrogen_production_from_the_thermal_splitting_of_methane_in_a_high_temperature_

23 www.hydrogensolar.com.

24 https://www.siemens-energy.com/global/en/offerings/renewable-energy/hydrogen-solutions.html.

25 https://www.energy.gov/eere/fuelcells/hydrogen-production-microbial-biomass-conversion.

26 https://en.wikipedia.org/wiki/Steam_reforming.

27 https://en.wikipedia.org/wiki/Pyrolysis#Methane_pyrolysis_for_hydrogen.

28 BASF. (2020). *BASF Researchers Working on Fundamentally New, Low-carbon Production Processes, Methane Pyrolysis*. United States Sustainability. Retrieved 19 October 2020.

29 https://en.wikipedia.org/wiki/Hydrogen_production#grey.

30 Jacob-Lopes, E., Manzoni Maroneze, M., Isabel Queiroz, M., and Queiroz Zepka, L. (2020). *Handbook of Microalgae-Based Processes and Products*. Academic Press.

31 (2020). *Fundamentals and Advances in Energy, Food, Feed, Fertilizer, and Bioactive Compounds*, 771–798.

32 https://www.energy.gov/sites/default/files/2021-09/h2-shot-summit-panel2-plasma-technologies.pdf.
33 Budhraja, N., Pal, A., and Mishra, R.S. (2023). Plasma reforming for hydrogen production: Pathways, reactors and storage. *International Journal of Hydrogen Energy* 48 (7): 2467–2482.
34 https://en.wikipedia.org/wiki/High-temperature_electrolysis.
35 https://www.sciencedirect.com/science/article/pii/S2590174520300155.
36 https://www.sciencedirect.com/science/article/abs/pii/S2542435123004968.

Chapter 07

1 H2 Tools. Refinery Hydrogen Production Capacities by Country. https://h2tools.org/node/820.
2 MEPC.1/Circ (2018). *878 Guidance on the Development of a Ship Implementation Plan for the Consistent Implementation of the 0.50% Sulfur Limit Under MARPOL Annex VI*. International Marine Organisation.
3 Duckett, A. (2018). Green steel process given go ahead for testing at pilot scale. *The Chemical Engineer*.
4 Penn State (2018). Carbon dioxide-to-methanol process improved by catalyst. *Science Daily*.
5 Hutchings, G.J. (2016). *New Catalyst Makes Hydrogen Peroxide Accessible to Developing World*. Lehigh University.
6 https://www.thechemicalengineer.com/features/uses-of-hydrogen-in-industry.
7 https://www.essentialchemicalindustry.org/chemicals/hydrogen.html.
8 https://www.thechemicalengineer.com/tags/clean-energy-sig-the-hydrogen-economy.
9 Adamson, K., Baker, A., Jollie Large, D., and Crawley, G. (n.d.). Fuel cell today. *Market Surveys*. http://www.fuelcelltoday.com/FuelCellToday/Analysis/Analysis External/SurveysHome/0,4328,,00.html.
10 http://en.wikipedia.org/wiki/Molten-carbonate_fuel_cell.
11 http://en.wikipedia.org/wiki/Phosphoric-acid_fuel_cells.
12 U.S. Fuel Cell Council. (2003). Fuel cell vehicle world survey. www.usfcc.com.
13 AE J2572 FCEV Standard.
14 http://www.gm.com/company/careers/career_paths/rnd/prj_well_to_wheel.html.
15 http://www.transportation.anl.gov/transtech/v2n2/well-to-wheel.html.
16 Wentz, W., Myose, R., and Mohamed, A. (2005). Hydrogen-fueled general aviation airplanes. In: *AIAA 5th Aviation, Integration, and Operations Conference (ATIO). September 2005*. Arlington, Virginia: AIAA 7324.
17 Fuel cell locomotive for military and commercial railways. Fuel 2003. October/November. http://Cell-Magazine.com.

18 http://www.cellular-news.com/story/18289.php.
19 http://www.mtimicrofuelcells.com.
20 Dokupil, M., Spitta, C., Mathiak, J. et al. (2006). Compact propane fuel processor for auxiliary power unit application. *Journal of Power Sources* 157 (2): 906–913.
21 de Bruijn, F. (2005). The current status of fuel cell technology for mobile and stationary applications. *Green Chemistry* 7 (3): 132–150.
22 http://www.plugpower.com/products/overview.cfm.
23 St-Pierre, J. and Jia, N. (2002). Successful demonstration of Ballard PEMFCs for space shuttle applications. *Journal of New Materials for Electrochemical Systems* 5 (4): 263–271.
24 McCurdy, K., Vasquez, A., and Bradley, K. (2003). Development of PEMFC systems for space power applications. *Fuel Cell Science, Engineering and Technology, International Conference on Fuel Cell Science, Engineering and Technology*. Rochester, NY, United States, April 21–23.
25 www.dodfuelcell.com.
26 www.aerovironment.com.
27 Rifkin, J. (2002). The Hydrogen Economy. Tarcher/Putnam.
28 https://www.researchgate.net/figure/Basic-structure-of-the-direct-methanol-fuel-cell-DMFC-system_fig1_339493770.
29 (2016). *Columbia Electronic Encyclopedia*, vol. 6. Columbia University Press.
30 *When Was the Battery Invented?* http://batteryuniversity.com/learn/article/.
31 *What Is a Lead Battery?* http://aboutbatteries.batterycouncil.org/.
32 *A Short Course on Charging Systems*. http://www.carparts.com/classroom/charging.htm.
33 Q & A: Car Batteries | Department of Physics | University of Illinois at https://van.physics.illinois.edu/qa/listing.php?id=567.
34 *Volt Electrical System*. https://en.wikipedia.org/wiki/42.
35 *Best Car Battery Reviews – Consumer Reports*. http://www.consumerreports.org/cro/car-batteries.
36 *Technical and Economic Feasibility of Applying Used EV Batteries in Stationary Applications*. https://www.researchgate.net/publication/236466652.
37 *Battery specs*. http://www.electricvehiclewiki.com/.
38 http://www.chargingchargers.com/tutorials/charging.html.
39 https://www.cargurus.com/Cars/2017-Toyota-RAV4-Hybrid-Overview-c26005.
40 *2011 Leaf First Responder's Guide*. http://www.nissanusa.com/. . ./leaf/2011/2011.
41 http://www.electricvehiclewiki.com/images/f/f7/2011-leaf-service-maintenance-guide.pdf.
42 https://www.nissanusa.com/ev/media/pdf/specs/FeaturesAndSpecs.pdf.
43 http://www.chevrolet.com/volt-electric-car.html.
44 *Energy Density*. https://en.wikipedia.org/wiki/.

45 *How Does a Lithium-Ion-Battery Work?* https://www.extremetech.com/extreme/184236.

46 https://en.wikipedia.org/wiki/Electric_vehicle_battery.

47 *Battery Electricity*. https://en.wikipedia.org/wiki/.

48 Bard, A.J. and Faulkner, L.R. (2001). *Electrochemical Methods*. New York: Wiley.

49 Norskov, K., Bligaard, T., Logadottir, A. et al. (2005). Trends in the exchange current for hydrogen evolution. *Journal of the Electrochemical Society* 152: J23–J26.

50 Adjemian, L.S.J., Srinivasan, S., Benzigen, J., and Bocarsly, A.B. (2002). Silicon oxide Nafion composite membranes for proton-exchange membrane fuel cell operation at 80–140°C. *Journal of the Electrochemical Society* 149: A256–A261.

51 Chunsheng, W. and Appleby, A.J. (2003). High-peak-power polymer electrolyte membrane fuel cells. *Journal of the Electrochemical Society* 150: A493–A498.

52 Meng, H. and Shen, P.K. (2005). Tungsten carbide nanocrystal promoted Pt/C electrocatalysts for oxygen reduction. *The Journal of Physical Chemistry* 109 (48): 22705–22709.

53 Hickner, M.A., Ghassemi, H., Kim, Y. et al. (2004). Alternative polymer systems for proton exchange membranes (PEMs). *Chemical Reviews* 104 (10): 4587–4611.

54 Uhling, H. and Revie, R.W. (1985). *Corrosion and Corrosion Control*. New York: Wiley.

55 Tang, H.L., Luo, Z.P., Pan, M. et al. (2005). Synthesis of platinum nanoparticles and then self-assembly on Nafion membrane to give a catalyst coated membrane. *Journal of Chemical Research* S7: 449–451.

56 Bevers, D., Wagner, N., and VonBradke, M. (1998). Innovative production procedure for low cost PEFC electrodes and electrode/membrane structures. *International Journal of Hydrogen Energy* 23: 57–63.

57 Gulla, A.F., Saha, M., Sudan, A. et al. (2005). Dual ion-beam-assisted deposition as a method to obtain low loading-high performance electrodes for PEMFCs. *Electrochemical and Solid-State Letters* 8 (10): A504–A508.

58 Ioroi, T. and Yasuda, K. (2005). Platinum-iridium alloys as oxygen reduction electrocatalysts for polymer electrolyte fuel cells. *Journal of Electrochemical Society* 152 (10): A1917–A1924.

59 Singhal, S.C. and Kendall, K. (2003). *High-temperature Solid Oxide Fuel Cells: Fundamentals, Design and Applications*. Japan: Elsevier Science.

60 Xue, X., Tang, J., Sammes, N., and Du, Y. (2005). Dynamic modeling of single tubular SOFC combining heat/mass transfer and electrochemical reaction effects. *Journal of Power Sources* 142: 211–222.

61 Koyama, O.M., Wen, C., Yamada, K., and Takahashi, H. (2003). Object-based modeling of SOFC system: dynamic behavior of micro-tube SOFC. *Journal of Power Sources* 118: 430–439.

62 Murthy, M., Esayian, M., Lee, W., and Van Zee, J.W. (2003). The effect of temperature and pressure on the performance of a PEMFC exposed to transient CO concentrations. *Journal of Electrochemical Society* 150: A29–A34.

63 Gu, T., Lee, W.K., Van Zee, J.W., and Murthy, M. (2004). Effect of reformate components on PEMFC performance: dilution and reverse water gas shift reaction. *Journal of Electrochemical Society* 151: A2100–A2105.
64 Wagner, N. and Schultze, M. (2003). Change of electrochemical impedance spectra during CO poisoning of the Pt and Pt–Ru anodes in a membrane fuel cell (PEFC). *Electrochimica Acta* 48: 3899–3907.
65 Ye, T.Y., Soo, K.H., Panakattu, K. et al. (2002). An NMR investigation of CO tolerance in a Pt/Ru fuel cell catalyst. *Journal of the American Chemical Society* 124: 468.
66 Yoshida, K., Ishida, M., and Okada, T. (2005). Tolerance to CO poisoning on fluorinated Pt catalysts for the anode of polymer electrolyte fuel cells. *Electrochemistry* 73: 298–300.
67 Baturina, O., Aubuchon, S.R., and Wynne, K.J. (2005). Thermal stability in air of Pt/C catalysts and PEM fuel cell catalyst layers. *Preprints of American Chemical Society, Division of Fuel Chemistry* 50 (2): 480–481.
68 Dhar, H.P., Kush, A.K., Patel, D.N., and Christner, L.G. (1986). Performance study of a fuel cell Pt-on-C anode in poisoning. *Proceedings - Electrochemical Society* 86 (12): 284–297.
69 Tewari, A., Sambhy, V., Urquidi MacDonald, M., and Sen, A. (2006). Quantificatio n of carbon dioxide poisoning in air breathing alkaline fuel cells. *Journal of Power Sources* 153: 1–10.
70 Atkinson, A., Barnett, S., Gorte, R.J. et al. (2004). Advanced anodes for high-temperature fuel cells. *Nature Materials* 3: 17–27.
71 Boennemann, H., Endruschat, U., Hormes, J. et al. (2004). Activation of Colloidal PtRu Fuel Cell Catalysts via a Thermal "Conditioning Process". *Fuel Cells* 4: 2970308. https://onlinelibrary.wiley.com/doi/10.1002/fuce.200400032
72 Komatsu, T., Arai, H., Chiba, R. et al. (2006). Cr Poisoning Suppression in Solid Oxide Fuel Cells Using LaNi (Fe) O3 Electrodes. *Electrochemical and Solid-State Letters* 9: A9–A12.
73 *Fuel Cell Handbook*, 5e (chapter 5). EG&G Services Parsons, Inc., Science Applications International Corporation Under Contract No. DE-AM26-9FT40575.

Index

Note: *Italicized* and **bold** page numbers refer to figures and tables, respectively.

b

c

d

g

r